IT MUST BE BEAUTIFUL TO BE FINISHED

A Memoir of My Body

KATE GIES

PUBLISHED BY SCRIBNER CANADA

New York   Amsterdam/Antwerp   London
Toronto   Sydney   New Delhi

An Imprint of Simon & Schuster, LLC
166 King Street East, Suite 300
Toronto, Ontario M5A 1J3

This Scribner Canada edition February 2025

SCRIBNER CANADA and colophon are registered trademarks of Simon & Schuster, LLC

For information about special discounts for bulk purchases, please contact Simon & Schuster Special Sales at 1-800-268-3216 or CustomerService@simonandschuster.ca.

Interior design by Wendy Blum

Manufactured in the United States of America

1   3   5   7   9   10   8   6   4   2

Library and Archives Canada Cataloguing in Publication

Title: It must be beautiful to be finished : a memoir of my body / by Kate Gies.
Names: Gies, Kate, author.
Description: Scribner Canada edition.
Identifiers: Canadiana (print) 20230591736 | Canadiana (ebook) 20230591817 |
ISBN 9781668051054 (hardcover) | ISBN 9781668051061 (EPUB)
Subjects: LCSH: Gies, Kate. | LCSH: Gies, Kate—Health. | LCSH: Gies, Kate—Mental health. | LCSH: Ear, External—Abnormalities—Patients—Canada—Biography. | LCSH: Body image—Social aspects. LCSH: Body image—Psychological aspects. | LCSH: Aesthetics—Social aspects. | LCGFT: Autobiographies.
Classification: LCC RF187 .G54 2025 | DDC 362.1978/10092—dc23

ISBN 978-1-6680-5105-4
ISBN 978-1-6680-5106-1 (ebook)

For my parents, Susan & Ken Gies.
For my friends on 7C.

An ear can break a human heart
As quickly as a spear,
We wish the ear had not a heart
So dangerously near.

Emily Dickinson

Table of Contents

*IT MUST BE
BEAUTIFUL
TO BE
FINISHED*

Author's Note

IT'S HARD TO PULL A LIFE INTO PAGES. SOME NAMES AND identifying features have been changed to protect the privacy of the people involved. Some characters have been made composites, and some events have been condensed. What follows is a work of nonfiction and a work of recollection. I've done my best to be as faithful to my experiences as my memory allows.

This is the story of a body that splits. The story of a shape-shifter, a work in progress, a body reaching for completion.

This story itself splits. I write to collect the pieces, to stitch them into something larger. To shore against my wreckages.

I write this body into wholeness.

I write to make this body my home.

Prologue:
The Lesion

THE DERMATOLOGIST SCANS MY HALF-NAKED BODY FOR suspicious moles. He asks me to pull my hair into a high ponytail so he can check my neck and ears. I am exposed. When he asks me about the right ear, his breath warming my neck, I say, "Just some surgeries." When he notices an abnormal lesion on the lobe, I say, "The skin is just weird there."

He insists on a biopsy, and I lie down on the examination table, head twisted to the left, right ear up—a configuration my body hasn't assumed in nearly twenty years. Panic thunders through me as he inserts the freezing needle. My stomach presses into my spine and I lose my breath. Despite the freezing, I feel him scraping the skin; I feel it in the back of my throat, in the spaces between my ribs, in the arches of my feet. I'm in my thirties, but for this moment, I am a child.

When I leave the dermatologist's office, I'm shaking. My whole body is flamed and raw. Decades-old scars, marking where the skin was split and resewn, reawaken. And the ear, an agitated heart, thumps and screams.

Later, my words to another doctor: *I can't live in this body anymore.*

I'm just as surprised by the words as my general practitioner sitting across from me. I'm surprised by their audacity, but also their accuracy. A truth teething inside me for decades has finally found its bite.

I can't live in this body anymore.

This body has never felt like a home. Not my home anyway. It's a body that's been given away many times. First, to hands trained to cut. Then, to any hands that would have it.

My doctor's hair is piled into a bun, and I note a slight raise of her ears as her face tenses into a grimace.

I can't live in this body anymore.

This is not a suicide thought, as my doctor believes. It's a simple desire to no longer be trapped in bone and skin. A desire, perhaps, to be disembodied. Floating without shape. Too slippery for fingers to grab, for scalpels to penetrate. Too ethereal to parse, to catalog, to own.

Part One

PREOPERATIVE

*A fiction is produced, a fiction that is a
projected image of the body.*

Elaine Scarry

IT STARTS WITH A BIRTH, A STORY TOLD IN FRAGMENTS.

Kingston, Ontario.

Me coming out the way many babies do: wet and screaming.

The doctors not letting my mother hold me.

The doctors taking me away.

Hours passing, my mother asking for me, a growing knot in her chest.

Then me, bound head to toe in a blanket, handed to her.

My mother noticing the deep purple of my eyes, not the barren space on the right side of my head: the space where an ear was supposed to be.

When the obstetrician told my parents about the missing ear, he asked my mother if she drank during her pregnancy. Although the answer was no, she stung with guilt. Was it something else she did or didn't do? Was it the time she slipped on the rain-slicked driveway? Was it the drug she took, prescribed for nausea? Was it something she ate that she wasn't supposed to?

I was placed in the neonatal intensive care unit for four days. The doctors didn't know where else to put me. None of them had seen a baby with a missing ear.

When my parents visited me, questions sparked: What would my life look like? Would people stare? Would kids tease? Would I have friends? Would I hate myself?

Later, a plastic surgeon knocked on my mother's hospital room door. He told my parents he'd seen a missing ear before, and that although I would never hear out of my right side, he could fix the physical deformity. He could make me look normal. Whole.

IT WAS A SMALL PEBBLE ON A LANDSCAPE OF SKIN—SMOOTH with textured edges, like half-chewed gum. The intricate slopes and folds of an outer ear missing.

Atresia anotia is the medical name attached to a missing ear. Atresia is any situation in which an orifice of the body is closed off or missing. *Anotia* refers specifically to the ear. The doctors' best guess was a cyst on my mother's ovary (that eventually grew to the size of a grapefruit) pushed against the side of my head during gestation, preventing the ear from forming, skewing the bones and nerves on the right side of my face.

The doctors gave me the name Atresia Anotia shortly after my parents gave me the name Katharine. With this new name came a set of instructions for what to do with my body.

I DON'T KNOW WHEN I BECAME AWARE OF THE MISSING EAR. I must have been two or three. The only memory I have is so fragile that when I dig into it, it breaks apart like tissue paper in water. It's a memory of my mother telling me I'd soon be going to a hospital so doctors could make me an ear. I remember only the edges: the hug she gave me in front of the kitchen sink; the sun through the blinds throwing diamonds on the yellowed floor; the foreign word *operation*—how I tried it in my mouth, but couldn't properly separate the syllables.

I remember my mother's candied eyes as she pulled me in. Then the hug. How she dug herself into me and I had trouble catching my breath with my face pushed into her stomach. How this hug worried me more than the word *operation*. I asked her if it was going to hurt, and I remember her answer. It was the same one she'd give when I asked before each of the many hospital visits that followed: "A little."

Me and Mom before the surgeries

MY PLASTIC SURGEON REMINDED ME OF AN OWL WITH HIS large alert eyes and tiny slit for a mouth.

"Here, touch it," Dr. Winston said, handing me a small piece of plastic. "That's it. That's going to be your ear."

The ear looked like a crooked C made of hardened milk. It felt small and smooth in my hands.

"It doesn't look like an ear," I said.

"Here." He took the plastic from me, pressed it into the right side of my head. "Hold it here."

My fingers replaced his.

He grabbed a hand mirror from his desk. "See?"

I did see. Against my head, the plastic ear took the shape of a real one; when I squinted, the whiteness blurred with the pink of my skin. In the mirror, I had two ears. In the mirror, I was whole.

There, in Dr. Winston's office, I became two bodies. I was the girl in the mirror and the girl looking into the mirror. The after and the before. The finished, the unfinished. The Kate I was supposed to be and the Kate I was.

Dr. Winston slipped the mirror back into his desk drawer. When he removed the ear from my head, I felt the pull of its loss.

"Once we slip it in," he said, "no one will ever have to know you didn't have an ear. It'll be like nothing was ever missing." He looked to my parents and nodded.

"How's it gonna get inside me?" I asked.

The adults laughed.

Dr. Winston described his plan. In the first operation, he would take skin from my buttocks and sew it to the side of my head. In the second operation, he would slip the plastic ear underneath.

"And then I'll have an ear there," I said. "Just like everyone else." I knew this was the important part.

"Yes, an ear like everyone else."

It would be only a couple of surgeries. Everything would be fixed and tidied up before I began school. Then my life with the body I was supposed to have could start. My mom and dad smiled. Dr. Winston smiled. I smiled, too.

Part Two

OPERATIVE

Liminal. A threshold. My body between worlds.

Terry Tempest Williams

One

I WAS PUSHED INTO A WORLD OF GREEN. THE MONSTER-HUM of machines grew louder. The green people closed in. *Katharine*, they said, their masks ballooning, *slide over to the table.* Hands raised me by my shoulders, my feet. The table was hard at my back. I sat up. *Lie back on the circle pillow. Look, it's like a doughnut!* It wasn't like that. The lights above turned my eyes into bruises. A cry rattled in my throat, staining the air. My hospital gown was pulled to my stomach. Hands all over me, planting orange circles on my neck, my chest. *Like stickers.* They weren't like that. A screen lit up with squiggles and lines. *That's your heart—look how fast it's going!* I was cold. A pinprick in my hand. I shut my eyes tight. Lips tingly. Rust in my mouth. I opened my eyes. The green people above me grew taller and taller like muscly weeds. My head floated away from my neck. I reached to grab it, but my arms didn't work. *Help*, I tried to shout, but my mouth didn't work. *Gooood, Kath-ar-ine.* Too much air in my name. Sinking. Voices above, stretched and broken. Everything far, but close. Like. Like in a. Bubble.

I REMEMBER FOLLOWING MY OMA DOWN TO THE BACK-woods behind our rented cottage, my little hand in hers. She had something to show me, something she knew I'd like. We came to a clearing, and there in front of us was a fallen tree, trunk horizontal, branches cracked and broken. The massive roots were exposed and vulnerable, reaching for the sky. I remember thinking how magical it was, a symphony of tangles and knots, the rootlets like delicate fingers, the sun decorating the rotting bark with puddles of gold. The more I looked at the tree, the more I became it. Its branches, my bones; its bark, my skin; its roots, my feet and hands. Stretching my fingers to the sky, I felt as big and magnificent as the tree. This was my body before surgery.

I remember gathering pillows on the living room floor. The Carpenters were on the record player, and Karen Carpenter's voice made the whole room glow. I was wearing an old yellow bedsheet as a cape, and my hair was in high pigtails. I kicked, I twirled, I bounced, my cape whirling about like a yellow wind. I climbed up on the couch, put my hands on my hips. I was a superhero. Indestructible. I jumped into the pillows. Softness into softness. When my mom came in to see what the ruckus was about, I was rolling in pillows. She took my hands in hers, lifted me into the air so I could touch the ceiling. We fell into the pillows, the music alive in our chests. This was my body before surgery.

I remember a beach, my feet like pale shells in the water, kicking up little storms of sand as I moved. Minnows darted and brushed against me. I stood still to feel their slippery tails on my legs. Back on the shore, I noticed a black pouch on my big toe. I tried to swat it away, but it stayed stuck. I ran to my mom and pointed at the thing. "A bloodsucker," she said. *Bloodsucker?* Horrified, I started kicking the air, trying to whip it away. It stayed stuck. My mom went to find salt, but I couldn't wait one more second with this thing sucking my blood. I dug my fingers into my toe and ripped the pouch off. Blood snaked from the creature as it wriggled in the sand. Blood snaked from my toe. I didn't care—the toe was mine again. This was my body before surgery.

THE ROOM WAS FUZZY. BEEPING AND FOOTSTEPS. I PEEKED through the bars of my bed and saw a fuzzy clock floating on a wall. I shut my eyes and opened them again. The fuzz was gone, and I saw other beds with bars on them, each with a kid trapped inside. I looked for my mom, but I couldn't find her. *Mom?* My words didn't work. The fuzz was in my mouth. I stuck my hand past my lips to pull it out. Nothing. I tried again, digging my fingers into the insides of my cheeks. Nothing. I petted my tongue; it was rough and sticky. Every time I took a breath, my bottom stung like I was being spanked. A nurse looked down at me. *Katharine, dear, you're awake,* she said. I didn't know her. *It's okay, dear. Everything's fine.* Everything was not fine. There were tubes coming out of me. Out of my foot. Out of my neck. The boy next to me howled. I tried to howl, too. Nothing came out. I scrunched up my face to cry, but tears wouldn't come. *It's okay, dear, you're just confused. Your operation is over.* The nurse put her hand on my arm and tears finally came to my face. *Mommmm!* My voice didn't sound like it belonged to me. *You'll see her soon*, the nurse said. *Just rest for now.* I tried to sit up and pain ripped through my backside. My head fattened with sick and I vomited yellow slime in my bed. The nurse brought over a tub that looked like a giant bean. *In here, dear.* I threw up again.

"THEY LET ME INTO THE RECOVERY ROOM AFTER THAT first surgery," my mother tells me, years later.

"I don't remember," I say.

"You kept calling out for me, you were so distraught. When I saw you, they had you on your side; you were crumpled into yourself. I'd never seen you so upset. I thought, *Oh my God, what did they do.*"

I try to picture myself as my mother saw me—small, terrified, curled into myself—but all I can conjure is a cartoony version of some other child.

"Dr. Winston didn't tell us how much skin he was going to take from your buttocks. I thought it would be a tidy incision, but it was a big, square wound," my mother says. "I was horrified."

MY COUSIN LUKE LIVED AROUND THE CORNER FROM US. He was older by three years and was always teaching me new things about the world, like how to make farts with my armpit and how to pump my legs to make me go higher on the swings.

A week after my first operation, he showed up at my house with a bag of marbles. When I came to greet him at the door, my newly carved behind making me slow and crab-like, he looked afraid—like if he touched me, I'd crumble like a sandcastle.

"Want to play?" he asked, holding up the marbles.

I looked to my mom. She nodded. "Go ahead, Bunny."

I followed him to the backyard. It was the first time I'd been outside since getting out of the hospital, and the smell of dirt and damp leaves was sweeter than I remembered. I squinted in the afternoon sun.

Luke hollowed out a small cave of dirt and showed me how to shape my fingers into a scoop to push the marbles into it. He told me to crouch to the ground.

When I tried to bend down, a sudden pain shot into my buttocks like the claw of a lion. I yelped and stood back up. It felt like I had split open. We looked at each other, momentarily terrified. What was this new thing happening to me? What was I turning into?

ON MY STOMACH, WITH NO PANTS AND NO UNDERWEAR.
Two nurses looked at the four-inch-square wound Dr. Winston had cut into my buttocks during my operation. I turned back to see what they were going to do but was told to be still. Swishing, then tugging. Hard. They were ripping the skin off. A pain so sharp my whole body twitched. Blue blotches flashed in my eyes, my stomach squeezed up to my chest. I heard humming and it was getting louder. It scared me. Then I realized it was coming from me and that scared me more. A scream cracked through from a place in me I didn't know existed.

"That's enough," I heard my mom say.

They didn't stop. They had to remove the bandages today, they said. It was in my chart.

The ripping continued. I soon became hoarse, my screams mostly wind. I kept at it anyway.

"Stop it!" my mom said. "Stop it now! I mean it. Call Dr. Winston."

They stopped and Mom put her hand on my back. Safe for now.

When they called Dr. Winston, they learned they weren't supposed to remove the inner layer of the bandage. That layer was supposed to fuse with the wounded skin, and eventually melt into it. But my screams couldn't stop them. Only Dr. Winston's words could stop them.

Two

WHEN I WAS FIVE, I LEARNED HOW TO SOMERSAULT. BEND, tuck head to chest, push forward, and roll! Like a wheel! I loved somersaulting so much, I stopped walking in my house. I somersaulted to the bathroom, starting as soon as I recognized I had to pee so I would make it on time. I somersaulted to the kitchen when called for dinner, holding up the family meal. I somersaulted to my room at bedtime. After a few days, the back of my head, where my body met the parquet, became sore. This wasn't enough to stop me. The fun of somersaulting far outweighed the pain of it.

What stopped me was my second operation.

The plastic ear was inserted under the grafted skin. It felt pointy and restless, not quite finding its place. Whenever I nodded or shook my head, a dull ache rang from ear to cheek. Whenever I rolled my head to the right in my sleep, the ear dug at my skull and I'd wake with a start. I couldn't somersault anymore. But when I imagined the ear, pink and perky under the bandage layers, I felt fuller. More real.

A FEW DAYS AFTER MY OPERATION, I LOOKED OUT MY front window to see Luke and the three boys from across the street getting into my neighbor's red car.

I shuffled outside and stood on my driveway.

"Where are you going?" I yelled. "Can I come?"

Luke and the boys shook their heads no. I looked to the neighborhood boys' dad in the driver's seat. He shook his head, too. "You better get back inside so your mom doesn't worry, dear."

The car took off, and I stood on the side of the road, bottom lip out, arms folded tight. I wondered if this was about not having a dink. I knew boys had dinks and girls had fou-fous. Dinks seemed more fun because they could make yellow pictures in the snow, and Luke once told me that he could stretch his dink backwards and touch his bum with it.

I grabbed a handful of dirt and threw it at the car as it turned to another street. Then I realized that with the boys gone, I could have the fort all to myself.

The fort was a wooden palace in my neighbor's backyard—a big cube with triangle peepholes carved out like eyes. It was a place where cops and robbers battled. Where knights killed kings, and wizards made potions out of mud and berries and pee. With the boys gone, I could be the complete boss of it.

I crept over to my neighbor's lawn, mad at my sore body for not being able to go as fast as I wanted. I started singing the song I sang when I was about to do something important.

"B-I-N-G-O! B-I-N-G-O! B-I-N-G-O! And Bingo was his name-O!"

When I arrived, the fort looked sad, its peepholes empty, its door sighing. I'd never been in the fort alone. Inside was all shadows. I stood on the wooden bench, stretched to tippy-toes to look out one of the peepholes. All was clear. I was the boss now. No one could come in, and no one could make me do anything I didn't want to. I waited for the red car to come back, ready to defend my new space. The car didn't come.

By the time I waddled back over to my own yard, my mom was waiting for me with her worried face on.

"Where'd you go?" she asked.

"They left me," I said, throwing my hands in the air.

"I couldn't find you. You were in the living room, and then poof!" My mom's fingers stretched into stars. "You were gone. Don't do that again. We have enough to worry about, Bunny."

I followed my mom back into the house and sat by the front window and waited. And waited. I stiffened as the red car pulled into my neighbor's driveway. The boys spilled out of the car and came straight over to my house.

"Kate! Kate! Come out!" Their voices bounced off the window like Ping-Pong balls.

I sat frozen, doing my meanest mad face.

"We got something for you!"

I unfroze.

I stepped out onto my lawn, completely forgetting about my mom and her worried face. Luke handed me a little pink package.

"Open it!" he said.

I tore through the paper, letting it fall in pink snowflakes. "Ohhhh."

It was a white satin pouch no bigger than my hand. On the front, a one-legged bird made of pink beads sparkled in the sunlight.

"A flamingo!" I said.

I'd never seen a real flamingo, but I was fascinated by the plastic ones that sometimes showed up with their pointed pink beaks and single legs on neighbors' lawns. I assumed flamingos were born with one leg just as I was born with one ear. I believed that if I were an animal, I'd be a flamingo.

"There's more," Luke said. "Open the pouch."

Inside was a tiny ballerina made of plastic. Her tutu was white lace and her hair was swept up into a high bun.

I looked up to the boys' faces shining at me and I felt warm all over.

EVERY SATURDAY MORNING, LUKE AND I CLIMBED INTO THE back of my dad's yellow Plymouth Duster and headed down to karate class in the gym of the psychiatric hospital where my dad worked. My dad was a karate instructor in his free time and insisted that both Luke and I take lessons, even though it meant missing our beloved Saturday morning cartoons. At five, I was the youngest in the class and one of the only girls.

The ride to the gym was my favorite part. The engine of the Duster was loud and vibrated my bones. We bounced in the black leather seats whenever we hit a bump.

In class, we punched and kicked and shouted in Japanese. I told my dad once that I wanted to take ballet lessons instead, but he told me ballerinas couldn't protect themselves like karate kids.

"You need to know how to protect yourself," he'd say. "You never know when or where you might be attacked."

After my operation, I didn't have to participate in karate class. I climbed up on the stage at the far end of the gym and played in the curtains. While people below me kicked, punched, and shouted in Japanese, I imagined my hair swept up in a tight bun instead of tangled in bandages. I imagined a white lace tutu, and two perfect ears shining from the sides of my head. I grabbed hold of the velvet and twirled until I was fully wrapped in curtain, like a ballerina in a cocoon waiting for her butterfly moment.

THE MORNING THE BANDAGES WERE SET TO BE REMOVED, it felt like a hummingbird was zooming in my belly. I couldn't wait to meet my new ear. I asked my mom if I could wear my yellow bathing suit with the rainbow cat on the front because it was my favorite thing in the whole world. She told me I had to wear real clothes. We settled on a brown corduroy dress with my bathing suit underneath.

"You ready to see the ear?" Dr. Winston asked when we arrived in his office.

"Yes!"

He patted his hand on the examination table, inviting me to sit. My buttocks were still sore to sit on, so I was slow to sink down next to him.

As he peeled the bandage from my head, I felt the sting of individual hairs pulled out at the root. Fresh wind on damp, newly exposed skin.

Dr. Winston handed me the mirror. When I looked into it, I jerked back in horror. Instead of the pink, perky ear I'd imagined, I saw a red, puffy mushroom with lines of wiry blue stitches. It looked like a Frankenstein ear. I didn't know how to run away from a monster gripped to my own head.

"It'll be better in a couple of weeks, once the swelling goes down," Dr. Winston said. "Trust me, you'll be loving that ear in no time."

I believed him.

THE LESION

In my thirties, three days after the dermatologist scraped for a biopsy sample, the man-made lobe continues to pump and throb. The ear is a drum, its beat thumping across my face, thrumming down my neck. My whole head becomes tender to touch. The right side of my face feels swollen. My neck stiffens, refuses to turn. When a strand of my hair lightly brushes across the ear, I'm overcome with nausea.

It's a familiar feeling—old roads remapping on the nerves, muscles, and veins of my body.

A lost circuitry relit.

Three

TWO WEEKS AFTER THE PLASTIC WAS INSERTED, THE EAR was still red and puffed. Every morning I'd squint into the mirror, trying to conjure the girl I saw in Dr. Winston's office when he held the plastic ear to my head all those months ago. I couldn't see her. Instead, it was just me with an angry red creature clinging to my head.

DR. WINSTON PRESSED HIS HANDS INTO THE EAR. A SHOCK sparked down my neck.

"Hmm," he said.

"What? What is it?" My mother's words were wavy with panic.

"Nothing to worry about." His hands continued to push into the ear.

I squirmed like a bird stuck in tar.

"Still now, stay still," he said.

I made my body as stiff as I could, trying hard to still myself. I counted backwards from ten over and over until Dr. Winston released his hands and I could breathe again.

AFTER MY VISIT WITH DR. WINSTON, LUKE CAME OVER TO cheer me up. We played on the swings in my backyard. Two bad guys in dark suits and grizzled beards were chasing us, wanting to snatch our bodies and make us into stew. But we were swinging up and away, faster and faster, the bad guys farther and farther below us. The swing set creaked and squeaked. We were in the sky, we were in the clouds, but the clouds weren't clouds; they were big lumps of Jell-O of every color: Gold! Silver! Pink! Aqua! Peach! We grabbed fistfuls as we flew past, fingers forming beaks, and we gobbled it all down and it made us stronger and my ear stopped hurting and we went higher, above the clouds, and the sky turned the bright blue of Smurfs, and we were powerful, and we were free, and no one could touch us.

ONE MORNING, I NOTICED A SMALL TEAR IN THE EAR. AN edge of plastic emerged through the skin's surface like a toy shovel in the sand. I ran my finger along the ear, skin to plastic. Feeling to no feeling.

I didn't know that my body was rejecting the ear. The skin was pulling away, pushing the plastic out.

Dr. Winston sighed when he saw it. "We have to take it out."

Take it out? Despite the heat of the ear, I was suddenly cold. Take it out? Meaning I couldn't keep the ear? Who would I be without an ear? I felt myself flicker like a loose lightbulb.

My mom's face was falling. "Are you sure?"

He was sure.

Soon after, I was admitted for surgery to remove the plastic ear. The extra skin from my buttocks was rolled up on the side of my head like leftover dough. Although it felt peaceful without the hard C beneath the skin's surface, without the pulse of pain, I cried over this plastic that was supposed to make me more real.

NOT LONG AFTER THE OPERATIONS BEGAN, I BECAME
terrified of the dark—a fear that would follow me into my preteen
years. When the lights went out for bedtime, it felt like I was extin-
guished, a candlewick snuffed by a firm thumb. Alone in my room,
under the blanket of blackness, I couldn't separate myself from my
surroundings. Did I exist? I would slap my arms and legs to make
sure I was still there. Sometimes I would try to break the darkness
with my voice: "This is Kate! I am here!"

My parents got me a night-light. Still, going to sleep was no longer a comfort but a falling. I brought books into bed with me: the
Berenstain Bears, *The Paper Bag Princess*, and later, Ramona Quimby,
Superfudge, and *The Lion, the Witch and the Wardrobe*. The words
in the pages kept my mind off the shadow of sleep. Books took me
to worlds where ghosts turned out to be wind, where kids defeated
witches, where everything worked out okay in the end.

A COUPLE OF WEEKS AFTER THE PLASTIC WAS REMOVED, MY parents signed me up for ballet lessons. I was so excited I almost threw up.

On my first day of class, the little girls bunched in a corner and we all took turns leaping and skipping to the other side of the room. I got to go first because I was the new girl. My stomach whirled as I bounced across the wooden floor, my pink ballet slippers flying and crashing and flying again. I felt like bubbles, like air, like I could jump high enough to break through the roof. I laughed with delight, still feeling like I was flying when I reached the other corner and lined up for another go.

The girl behind me tapped my shoulder. When I turned around, she pointed to my head where the plastic ear used to be, now a gnarl of skin and fresh scars. "That looks really gross. And your mouth is really crooked when you laugh."

She waited for me to say something. I didn't know what to say because I knew she was right. A heaviness budded in my knees.

"Don't cry," she said. "Shhhh. Don't cry, I'll get in trouble."

I cried. She got in trouble.

THAT NIGHT, WE ATE DINNER LATER THAN USUAL BECAUSE of the ballet. Afterwards, I headed to the TV and turned it to Channel 2. Much to my surprise and horror, the *Polka Dot Door* was not on. I ran into the kitchen and yanked at my mom's pant leg.

"Mom, where's the *Polka Dot Door*?"

"It's over, Bunny. We were eating."

"But what about Marigold!" Marigold was my very favorite character and I looked forward to seeing her every night.

"You can watch it tomorrow."

"But I want to watch it today!"

"Well, you can't do that."

I spent the next half hour in near tantrum, lamenting the loss of missing the *Polka Dot Door*, paddling between my mom and dad to ask in desperation if there was any way I could watch it, until I got sent to my room.

Alone on my bedroom floor, I pulled out a piece of paper and a pencil and decided to write a note to explain my deep suffering for having missed my favorite show in the whole world. I couldn't yet put into words that what I was really upset about was my body failing me, making me forever gross. When I was done writing the note, I slipped it under my door into the hallway for my parents to find. On it were two words boldly printed in capital letters: "SHE DIDE."

Me and my Marigold doll

THE KATE I'M SUPPOSED TO BE WAS THE KATE I WOULD HAVE been if my body had held onto the plastic ear. Better yet, she was the Kate who was born with two proper ears. Around the age of six, I started to compare myself to her constantly. The Kate I'm Supposed to Be wore her hair in a ballerina bun, and no one pointed at her with disgust, and she had a perfect smile, and all the other ballerinas thought her jumps were graceful and beautiful, and everyone wanted to be her friend.

Four

"I HATE YOU," I SAID TO MY MOM, FISTS CLENCHED, FEET stomping hard into the tile of the mall floor.

She didn't stop moving. We walked past the craft store, and I saw sparkly balls and painted bowls and stacks of crayon-colored yarn in the window display. I needed to go in and get supplies for the shoebox house I was making for my plastic chipmunk. My mom had said we didn't have time. Too busy today; we were on a mission. I didn't know what a "mission" was, but it seemed like something horrible if it didn't include aisles of ribbons and buttons and colored pencils.

"We have to get Gramma a birthday card," my mom said. "There are so many to choose from. You can help me pick one out!"

She was trying to make it sound fun, even though I knew it wasn't going to be fun because cards weren't fun. They were the boring things that you had to pretend to enjoy before you were allowed to open your presents. I followed behind my mom slowly, milking my disappointment by shuffling my feet.

We entered the card store and I followed my mom to the birthday section.

"Look at all these lovely cards!" she said in her squeaky "I'm talking to a baby" voice.

I turned away from her with a huff, pretending to look at the cards on the other side of the aisle.

Something caught my eye: a card with a picture of a sad puppy on the front. Wrinkled with a pushed-in nose, the puppy stared at me with droopy, wet eyes. My chest lurched. What was wrong with the puppy? I wondered. Why was she so sad? Where was her mom?

Guilt crept in. I'd just told my mother I hated her, and this poor puppy maybe didn't even have one. I imagined the little puppy opening its little mush mouth and crying, *Mommy! Mommy! Where are youuuuu?*

I turned back to Mom and wrapped my arms around her middle.

"I love you," I whispered dramatically.

Her body stiffened. I looked up to see the pouchy face of an old man staring down at me.

I hadn't heard my mother when she told me she was going to another row. I hadn't heard her walk away and be replaced by this strange man.

I pulled my arms away from his waist, my face hot with embarrassment.

IN ADDITION TO HAVING NO OUTER EAR, I HAD NO HEARING on the right side. This meant sounds from that side were severely blunted, and often unheard. It also meant I couldn't figure out where sounds came from (a trick the brain needs two ears to do). As a young kid, I often ran in the wrong direction when cars honked at my back, sometimes running into a moving car as it tried to pass. If someone in my house called for me from a different room, I would sometimes have to search every room to find them. If I didn't sit at the front of the classroom, I'd miss large bits of instructions. Playing Marco Polo in a pool was not fun for anyone, because "Polo" would hit me at all angles and I'd end up being Marco indefinitely until we all gave up. Whenever we played telephone at school, I was the one who always broke the message.

Then the ear infections in my left ear started. I was five or six when they got bad. With the left side plugged up, and no hearing on the right side, everything around me became indecipherable. My hearing was reduced to whooshes and warbles.

I was split off from the world, sealed away in a glass jar.

ONE DAY IN GYM CLASS, DURING A BAD EAR INFECTION, OUR teacher stood in front of us giving directions. I looked around to the other kids to try to figure out what she was saying. The kids nodded and broke into two groups in the middle of the gym. Where was I supposed to go? I joined one group but was met with a wall of wagging heads and fingers pointing me to the other group. With hot cheeks, I turned around and joined them.

Our teacher came over and said something blurry. We formed a circle and my hands got snatched up by the kids on either side of me. Once we were all holding hands, my classmates started marching their feet and the circle rotated. As I marched along, I noticed my classmates' mouths opening and closing, eyebrows high, fuzzy sounds coming from their faces. They were singing. I widened and shut my mouth along with them to make it look like I was part of the group. Suddenly everyone was on the floor, and I was still standing, no longer attached to anyone's hand. I shot down quickly, hoping nobody had noticed.

Just as suddenly, we were up again, clasping hands and rotating once more. My teacher joined our circle this time, grabbing my hand and directing us all in the rotation. I watched closely for signs of when I was supposed to fall. When I noticed one of the girls in the circle lean forward, I quickly dropped to the floor. Too soon. With my hands still attached to a classmate on one side and my teacher on the other, I dragged them both down with me, creating a domino effect of falling kids. A mess of jumbled legs. Scrunched faces in my direction, heads shaking. I looked to my teacher. I could see up her skirt, and she was wearing beige underwear. She turned to me, her face red, her lips moving fast. I didn't know what she was saying, but I knew I was in trouble.

THE EAR INFECTIONS WORSENED. WORRIED THAT MY PERI-ods of near deafness would affect my learning, my parents consulted two separate specialists who told them that surgically inserting a plastic tube in my left ear to drain the fluid was the best way to go. It was risky, given it was my only hearing side, but continued ear infections could also be risky.

When I woke up after the operation, there were no bandages and no cuts on my body. Just a simple dull ache in my left ear.

A COUPLE OF DAYS LATER, I WAS BORED. EVEN THOUGH MY ear still ached, I was so restless that my mom told me I could call on my friend Lindsey from up the street. Before I went, I got the brilliant idea to get us Freezies from the basement.

In my excitement, I tripped on a step and tumbled down the stairs. The basement floor slapped me in the face. I lay there, dazed and foggy, rubbing my head. A sudden pain in my left ear, but it only lasted a split second. My legs felt like a baby deer's as I stood and wobbled to the freezer.

When I got to Lindsey's house, she was sitting on her lawn, staring at a wooden puzzle in front of her. I skipped over and placed a blue Freezie beside her lap. She didn't notice. She moved around puzzle pieces in the grass, trying to figure out how they went together. I sat down beside her and, without a word, began to move the pieces around with her.

"Don't help!" she said.

I removed my hands. I could see how the puzzle fit together and I was afraid my hands would try to help again, so I stuffed them under my knees. Then I felt a tickle in my left ear. I ignored it at first, pressing my hands hard into the backs of my knees. The tickle soon became a pounding itch, and I had no choice but to release one of my hands to scratch it. As I pulled the hand back, I saw it was slicked red. It took me a moment to realize the red was blood. My blood. My head started to crackle. Everything became a low hum.

I shot up and sprinted toward home without saying goodbye. As I ran, blood dribbled down my neck and the crackling in my head got worse. It sounded like the world was breaking apart.

My mom was in the kitchen. I stood in front of her, bloody-eared, and searched her face. Her mouth moved; no words came out. A whir escaped her lips. She was calling for my dad.

My mom and dad danced in front of me in frightful jerks, scrambling for keys, shoving feet into shoes. Amid the crackling, words came in and out of focus like heartbeats.

. . . her health card . . .

. . . I don't know . . .

. . . Bunny, can you . . .

We arrived at the front desk of the emergency room. The nurse waved us in when she saw the blood creeping down my neck.

Then I was lying left ear up on a stretcher with four men peering over me. One of the men bent down to me. "Hemmmimble ahahch uction."

A loud ring filled my head. They were sucking the blood out of my ear. Bloodsuckers. I shut my eyes very tight and made balls of my hands. I opened my eyes a sliver and saw my dad against the far wall. He was white and droopy. When he saw me looking at him, he tried to smile, but instead, he slid down the wall. A moment of relief when the men stopped the sucking and rushed over to my dad.

It was all the blood, he told me later. He fainted from all the blood.

AFTER THE EMERGENCY ROOM, MY MOM TOLD ME THAT THE tube they'd put in my ear to help me hear had ruptured my eardrum, which meant I couldn't go swimming that summer until it healed over.

None of us knew then that the eardrum would never heal over. None of us knew that the damage was permanent, that there were certain pitches of sound I would never hear again. That although I would be fitted for many earplugs over the years, none were foolproof. Swimming would always be a hazardous game of keeping my head above water.

There was talk, a few years after the rupture, of trying to fix the eardrum. But no one was willing to take the risk.

I SPENT MANY YEARS MISSING CLASSROOM DISCUSSIONS AS a student because I couldn't follow the popcorn flow of conversation. I became exhausted by my concentrated effort when communicating with soft talkers. I read lips. I nodded even when I missed parts of what people said. I accepted when someone behind me got mad when I didn't move to his "Excuse me." I automatically laughed at jokes I missed. I got lost in complicated, dialogue-heavy movies. I couldn't pick out songs playing in the background at restaurants. I didn't know the sound of my own footsteps. The sound of my breath.

ALTHOUGH I WAS DEAF ON ONE SIDE AND HAD A RUPTURED eardrum on the other, it wasn't until I started teaching at the local college, when it became crucial that I hear all lines of classroom discussions, that I considered myself hard of hearing.

My problem had always been perceived as the physical deformity of the missing ear. Its form was the problem, not its function: what I looked like, not the ease with which I took in and navigated the world.

THE LESION

A week after the dermatologist appointment, with my head still tender and flaring, my left ear starts to plug up. My world is now underwater. Sounds are rounded, muted, wrapped in fur. My voice is bubbled.

I take the subway to the walk-in clinic. I avoid eye contact so that I don't have to decipher the extra sounds of spontaneous conversation.

At the clinic, I sit low in my chair, eyes fixed on the hall where the nurse comes out, to make sure I don't miss when she calls me in.

In the examination room, I tell the doctor I'm having trouble hearing. He says something low and blurry.

I can't hear you.

He raises his voice and I focus on his lips. What's going on? he asks.

I think I might have an infection in my left ear.

He places the scope in my ear, says something that I don't hear. He pulls away. I focus on his lips, ask him to repeat what he said.

The eardrum is ruptured, he repeats in a forced tone.

I know that. It's been that way since I was five.

He wants to look at the other ear. Before I can stop him, he brushes my hair aside and pulls back when his ear scope meets skin instead of an ear canal.

I don't have a proper ear on that side, I say.

What happened there? He pulls back my hair again. I know I'm going to have to indulge his curiosity.

I recite the story of the ear and then brace for the questions I always get from doctors who don't know me.

DOCTOR: [question]
ME: Born that way.

DOCTOR: [question]
ME: Nope, no illegal drugs during the pregnancy.

DOCTOR: [question]
ME: No, I'm not interested in trying that.

He gives up on the questions about the right ear.

You should see your regular doctor about the left ear, he says. Because of the rupture, I'm not sure if it's infected. But I notice your mouth is crooked when you talk.

(This is the other thing that doctors like to bring up.)

I know.

It can be a sign of a stroke.

I'm not having a stroke.

Five

A FEW MONTHS AFTER THE EARDRUM RUPTURE, MY MOTHER and I were back in Dr. Winston's office. He wanted to try again: He wanted to insert another plastic ear on the right side. Mom wrapped her arms around me and squeezed tight. It was a new start, a new hope.

"We'll get an ear in you yet," he said, smiling at me as wide as his little mouth would allow.

After the operation, I was extra careful in my body. I walked slowly, and petted the ear through the bandage to make it feel comfortable inside me. I made sure not to mash it into my pillow when I slept. Every throb, every itch sent a flush of panic through me. I made deals with the ear. If it stayed in me today, I'd eat all of my broccoli to send it healthy food. If it stayed in me, I'd go to bed early to help it heal faster. I started giving myself rules to avoid bad luck. If I stepped over sidewalk cracks, the ear would stay in me. If I counted to seven every morning and every night in front of the bathroom mirror, the ear would stay in me. I made up a song to the melody of "Twinkle, Twinkle, Little Star" and sang to the ear every night at bedtime: *Hold on, hold on, little ear . . .*

After surgery

A FEW DAYS AFTER THE OPERATION, I SNUCK INTO MY PAR-ents' bedroom and grabbed a pair of clip-on earrings from my mother's jewelry box. Gold hearts dangling from a pearl stud.

Back in my bedroom, I clipped one to my left ear and did a twirl. The earring jingled and slapped my cheek glamorously. My right side was still bandaged up, so I clipped the other earring to the bandage, as close as I could get it to where an earlobe would be. In the mirror, I could see her: the Kate I'm Supposed to Be. I did another twirl. The earring fell from the bandage and plinked on the floor.

MY FATHER AND I WOULD SOMETIMES GET UP AT THE HINT of sunrise, pack some snacks, and take a walk in a wooded park, its trails arched with tree branches and carpeted with soft dirt. We'd look for little creatures—frogs, squirrels, birds—and discuss life. One time, we talked about the ear.

"It's going to be like a Corvette!" my dad said. "Corvettes are made out of the best plastic in the world. All the rest of us have these boring Oldsmobile ears. But you! They're building you your own fancy one-of-a-kind ear. With the best plastic in the world!"

I tried to match his excitement. A Corvette ear. One of a kind. Specially made. Although I knew my dad was trying to make me feel good, something dug at me: I didn't want a special, one-of-a-kind ear. I wanted an Oldsmobile ear.

WHEN THE BANDAGES WERE REMOVED THIS TIME, I DIDN'T want to look, afraid I'd discover the monster ear from last time. I closed my eyes and refused the mirror.

"Are you sure you don't want to see your new ear?" Dr. Winston asked.

"Yes." I kept my eyes squeezed shut. "I want to wait until it doesn't look gross."

"It doesn't look gross, Bunny," my mom said.

My curiosity quickly got the better of me. I softened my eyes and opened them. Dr. Winston placed the mirror in front of me.

I pulled back and let out a little shriek. The ear was the same monster from the last time the bandages came off.

"Now, come on, it doesn't look that bad." Dr. Winston was chuckling.

"Bunny, it really doesn't," my mom added.

It did look that bad. The same red mushroom and puckered lines of wiry blue stitches. I couldn't tell if Dr. Winston and my mom were trying to trick me into thinking it didn't look scary, the way adults sometimes told lies to make kids feel better, or whether both of them needed glasses.

School diary entry

DR. WINSTON GAVE MY MOTHER SPECIFIC INSTRUCTIONS for cleaning the ear. Mild soap on a facecloth, wiping in downward strokes. Lots of Polysporin on the incisions.

Three times a day, I would stand tight in the bathroom and pull my hair away from the ear as my mom wet and soaped a cloth.

"Ready?" she'd say.

"I'm just going to take some breaths first." I would take a breath and exhale loudly.

"Ready now?"

"No." I'd take another breath. The newborn ear felt raw and delicate against my head, and I was afraid the facecloth would scrape off the thin layer of skin holding it all together.

"Come on, Bunny, let's get it over with."

"Okay." I'd squeeze my shoulders into my neck.

She'd approach with the cloth.

"Ow ow ow ow ow!" I'd start even before the cloth reached me.

When the facecloth hit the ear, it felt like a branch of pine needles. I'd pull my lips under my teeth, bite down, and hum the theme song to the *Polka Dot Door*. The thirty seconds it likely took to wash the ear felt like hours. Just when I thought the ear would tear off, my mom would remove the cloth.

"Okay. The Polysporin now, then we're all done." She'd pull out a Q-tip and cover it with the clear, greasy goop.

"Ow ow ow ow ow!" The Q-tip didn't really hurt, though. Sometimes it actually felt soothing, like sinking into a warm bath.

"All done, Bunny."

I'd release my hand, and my hair would flop back over the ear like a shield.

IN THE CRACKS OF MY DRIVEWAY, ANTS WERE BUILDING TINY sand pyramids. I rolled up my sweatpants and knelt to get a closer look at the lines of black bodies carrying bits of sand on their backs. I ran a finger lightly along the edge of the new ear, checking for numb spots—a habit I'd formed in the weeks since the bandages were removed. The ear was less red than the first time, and although it still looked scary, Dr. Winston said it would probably stay in me.

"Katy!" My mom jutted out from the front door. "Time to get ready for ballet."

"Jus' a minute," I said and turned my head back to sand pyramids and ant lines.

"We're going to be late," she warned.

But I was busy with my ants.

I stuck my foot out in front of the ant line, and they started moving upward, their tiny bodies forming a dotted black arc across my running shoe.

"Kate!"

"Not right now, Mom!"

More ants climbed my shoe.

A shadow was forming. Mom was standing above me, car keys in her hand.

"We have to leave now. No more time to get ready."

I stared at her. "But what about my ballet suit?"

"No time, Bunny. You'll have to go as you are."

I arrived at ballet wearing my sweatpants and running shoes. My ballet teacher looked at me with a bent eyebrow. "Good afternoon, Miss Kate."

"Hi," I said, mouth in a tight frown. I joined the other girls sitting in a circle in the center of the room. They stared, but this wasn't an entirely new thing. They already thought I was weird because of the ear, and because I sometimes sang to myself when I danced.

"Class, today is a special day," my teacher said as she pulled out a large bag of pink clouds. "Your tutus for the recital are here!"

Little girls clapped with excited little hands.

The other girls gracefully slipped their tutus over their black bodysuits, while I could barely get my shoes through the leg holes. My tutu, stretched over sweatpants, shot out at odd angles from my waist. Stares and giggles then shot out from little ballerina heads.

In the mirror, the girls were a long line of black bodices and perky tutus, interrupted by me. Their hair was swept up into high buns, while mine was low and scraggly. The skin around the new ear was too sensitive for me to get my hair into a proper bun. Also, I wanted the ear hidden so that none of the girls would point out how gross it was.

"All right, girls, ready?" My ballet teacher stood in front of us. "First position, please."

We each pulled our feet into a V. The music started and we began our recital dance. Plié, plié, jeté, pirouette. My shoes squawked as I spun, and despite doing my best to pat down my pants, they flapped and billowed like sails on a windy day. My jumps landed in clumsy thwaps. Grease from the Polysporin made my hair stick to my face. All around me were miniature ballerinas, soft and graceful. The more I tried to match their elegance, the heavier, louder, and uglier I became.

"PASSING," WRITES SANDER GILMAN IN *MAKING THE BODY Beautiful*, "is a means of trying to gain control. It is a means of restoring not 'happiness,' but a sense of order in the world. We pass in order to regain control of ourselves and to efface that which is seen (we believe) as different, which marks us as visible in the world."

I knew I could never be like the other ballerina girls. They didn't seem to have to fight their bodies to be normal. They could pull back their hair into high buns without a second thought.

I was starting to realize that the surgeries couldn't make me normal—all they could do was make me *look* normal. Although the plastic ear could perform the appearance of a real ear, it wouldn't work like a real ear or feel like a real ear. It wouldn't change the fact I was born without an ear. The surgeries could help me pass as a two-eared girl, but, like a tutu shoved over sweatpants, they could not erase who I was: a fraud.

THE NIGHT OF THE BALLET RECITAL, WE WAITED IN THE wings at downtown Kingston's Grand Theatre—the biggest theater in the city. I'd been there once before, in the audience, to see the musical *Oliver!*. I never imagined that I would be part of a show at the Grand. From the wings, the stage looked much bigger than it did when I was in the audience. As we waited to perform, our cheeks and lips slicked with our mothers' blush and lipstick, we practiced our smiles. I rustled my hands in my tutu; it felt like a fishing net.

"All right," our teacher whispered to us, "you girls are up!"

When I got to the middle of the stage, I looked out to find my mom and dad. I couldn't see anything. A spotlight beamed on us so brightly that it washed away my sight and made my eyes hurt, like the lights of the operating rooms. When I closed my eyes, the lights still blazed under my eyelids.

The music started. It was very soft—much softer than when we practiced. I didn't understand that the speakers were facing outward, blasting the music into an audience I couldn't see. The music wasn't for us, it wasn't for me, and because of my bad hearing, I could barely catch it. I didn't know when I was supposed to start dancing. I looked to the other girls; they were already moving. I started moving with them, mimicking the gestures of the dance, but without the music alive in my body, it didn't feel real. I didn't feel real.

THE LESION

A week after the dermatologist and two days after the walk-in clinic, I see my doctor about the pain on the right side of my head and the plugged left ear. She examines me and says there's some slight swelling on the right side, likely due to the scraping for the biopsy. Nothing to worry about. She looks into my left ear. No infection. She tells me the muted hearing is likely a blockage in my Eustachian tube.

"Nothing can be done about it, it will clear on its own over the next few weeks," she says, having to repeat it several times for me to hear it. She doesn't seem concerned by how weeks of severely compromised hearing will affect my job, my relationships, my ability to negotiate even the simplest human interactions.

At home, I Google "Eustachian tube blockages and swelling," desperate for a quicker solution. The internet tells me that both can be due to "inflammation." The internet leads me to site after site about how to diminish inflammation. No gluten, no sugar, no caffeine, no soy, no dairy, no tomatoes, no fried food, no starch, no eggs, no beef, no pork, no processed food. A surge of optimism rushes through me. I can control this. I can purify my body.

I go through my cupboards and fridge and throw out all of the foods on the list, emptying most of my kitchen contents. I go to the health food store, buy organic spinach, lemons, gluten-free sprouted bread, and honey. In a matter of days, food becomes an obsession, a new way to exert control over my unpredictable body. It will be weeks, months, years of this, and my body will shrink closer and closer to its bones.

IT WAS A CRACK AT THE TIP OF THE EAR. SKIN SPLIT.

A month after Dr. Winston inserted the new plastic ear, my mom and I sat in his office, waiting for him to say something. He opened his mouth, then closed it. He did this a few more times before Mom spoke.

"Just say it," she said.

He scrunched his face up like a rotten plum. "It's infected. We have to take it out."

The news felt like a punch. I looked to my mom. Her face became sharp angles like she was going to cry, which made me want to cry.

No ear. It was over. I would never be complete. I would never come close to the Kate I'm Supposed to Be, and my mom would cry every time she looked at me.

"It didn't work," she said.

"I'm sorry. I don't think there's much more I can do."

"You're sorry. Three years. Four operations on that ear. Five, now that you have to remove it." She shook her head. "What a waste."

"You should take her to Toronto. There are surgeons there who specialize in ears." He pulled out a sheet of paper, wrote something down, and handed it to my mother. "They won't use plastic; they'll take bone and cartilage from her own body, so she'll be less likely to reject."

Cutting out my own bones? I immediately thought of a ghost story I'd once heard about a lady who was forced to eat herself, bones and all, until she was nothing. I looked to Mom. She stared at the floor.

Dr. Winston leaned forward. "I can understand your frustration but—"

"No. You can't." My mother pulled me out of my chair, my fingers tangled in hers, and we walked out the door.

Six

THE PLASTIC WAS REMOVED IN ANOTHER OPERATION. The skin graft rolled up once more. Red, crusted incisions like dried-out worms over a graveyard of used-up skin.

"I was heartbroken," my mother tells me, years later. "He'd promised an ear. He'd been so confident about it, so reassuring. And then it was over, and he said he couldn't do anything else. I was . . . I don't know . . ."

"Mad at him?" I ask.

"Not mad. He was such a nice man, Kate, he really was, and I could tell he was disappointed, too. But it was like he completely gave up on us after being so sure he could give you an ear. He made us all believe in him, in the ear, and then poof! Just like that, he dropped us. I felt defeated. It was so unfair, all those surgeries, and they amounted to nothing."

"Worse than nothing."

"Yes, I suppose that's true."

I FOUND THE MOUSE UNDER THE PINE TREE IN MY BACKYARD a few months after that operation. Her eyes were frosted gray, her nose whitened by snow, her tiny pink paws folded into fists. I crouched in closer, expecting a burst of movement from her. Nothing. When I took off my mitten and touched her, her body was hard and cold. "Hi," I whispered to her, "I will fix you." I gently scooped her into my hands and headed to the back door, slipping past Mom in the living room and climbing the stairs to my bedroom. I pulled some tissues from the box on my dresser, wrapped my mouse to her chin, and placed her inside the box. I marveled at the small, slivered teeth peeking from her mouth, the delicacy of her paper-thin ears, the shine of her whiskers. I sang to her. "Hold on, hold on, little mouse." If I could save this little body, everything would be okay. I stayed with her as the sun bled orange through the bedroom blinds, then disappeared into darkness. I stayed with her, stroking her whiskers, waiting for her to thaw into life.

THE LESION

I'm in a room, sitting in a chair across from a man who calls himself The Specialist. The room is white and impeccably clean; the ivory-tiled floor reflects the bright ceiling lights, creating splotches of illumination around my feet.

"How is your eating?" he asks.

"I'm being very careful."

The Specialist pulls out a scalpel. "I have something to show you."

He grabs my hand, turns it palm up, and places the point of the scalpel at my wrist. I try to pull my arm away, but he's too powerful. He sinks the blade into my skin. Blood spills from my wrist and dots tiny firecrackers onto the floor. He draws the scalpel upward toward my elbow. The skin breaks willingly.

He stares at the wound he's made. A thin rivulet of dark matter seeps out of it, like smoke escaping a fire. He asks me if I feel them. Before I can say *Feel what?*, there's wriggling against my bones. I look into the wound. A scream catches in my throat. Maggots, wiggling between my veins, squirming into muscle, their milky bodies shiny with blood.

The scream dislodges and pollutes the sterile white room, echoing off the walls.

I awake in darkness, my skin slick with sweat. My breath rapid. I slap my hand against my arms. No cuts, no blood.

But I can feel a hollowing. Millions of cells suiciding each second.

Seven

IN NATHANIEL HAWTHORNE'S SHORT STORY "THE BIRTH-mark," a scientist becomes obsessed with a small stain on his wife's cheek. When he asks if she's ever considered removing it, she tells him no—that she and many others have always viewed the mark as a charm. The scientist grows increasingly fraught when he looks at her. The stain, once small, becomes the only thing he can see. He dreams that he cuts the mark out with a blade and, realizing its roots sink all the way to her heart, cuts her heart out, too. The scientist tells his wife about his dream, and she, seeing the distress the mark is causing him, agrees to have it removed. The scientist devises a series of remedies: a magic flower, metal-plate portraits with the birthmark erased, fumes pumped into the air. As the experiments continue, both husband and wife become more and more obsessed with a pristine face. The wife sees him as noble for working so hard to achieve his ideal vision of her. At last, the scientist concocts an elixir that eradicates the thing for good, with only one major side effect: She dies.

"THE MOMENTUM OF PREVIOUS DECISIONS MAKES STOPPING difficult to consider as an option," writes medical sociologist Arthur Frank in his article "Emily's Scars." "Momentum reinforces the promise of a better life (a better self) if one more medical step is taken."

It didn't occur to any of us to give up on the Kate I'm Supposed to Be. We'd already risked so much. We couldn't go back to nothing. We wouldn't.

ONE MORNING WHEN I WAS SEVEN YEARS OLD, MY MOTHER stood in the doorway of my bedroom. "Wake up, Bunny, we're going to Toronto today. Operation time."

I closed my eyes quickly. If I was still asleep, I wouldn't have to listen to her.

"Kate, come on, time to get up."

I kept my eyes closed. She had warned me this was going to happen, that one day we'd be going up to Toronto. She never told me my operation dates because she didn't want me to spend weeks worrying about them. This kindness also meant the announcement that I was about to have an operation felt unreal, like a bad dream.

It had been almost a year since my last surgery. I was getting used to being a kid who didn't get cut into.

"Kate, come on," she said, tapping her fingernails on the wood of the door.

I opened my eyes. "But I can't go today. We're making candy apples at school."

"No, Bunny, you'll have to miss that. Today we're going up to Toronto."

"But I told Elena I'd teach her how to make paper frogs at recess today—I have to go, we made plans."

"Elena will understand. You can play with her when you get home."

"Does Dad know about this?"

Mom sighed. That's when I knew it was real. I could feel my heart in my neck.

"You can bring a toy if you like."

THE ONLY TOY I COULD THINK TO BRING WAS MY PUFFALUMP, Jim: a feather-light stuffed dog made of teal parachute material. He had a large pink gumball nose, and kind blue-threaded eyes. His marshmallow arms were much bigger than mine, and he gave the best hugs.

Jim became my comrade for many operations to come—he'd accompany me into every operating room in Toronto, and be lying beside me in every recovery room I woke up in.

When I find Jim in my parents' basement many years later, he'll be under a pile of old blankets, the weight of which will have flattened and condensed his innards. He'll look worn out, the parachute material hanging loose and wrinkled. He'll be smaller than I remembered. His eyes will seem vacant and cartoonish. An outline of translucent brown will stain his right arm and part of his belly. Iodine maybe? Vomit? Blood? His belly will be torn, his left leg attached by mere threads. A medical bracelet, an exact copy of the ones I wore on my wrists in Toronto, will still circle his right arm. The blue writing will have bled into blotches, making the words unreadable.

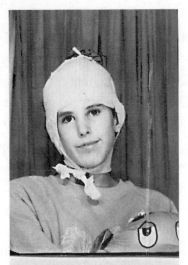
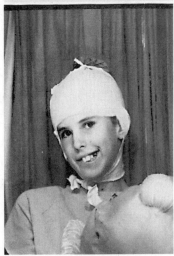

Me with Jim

MY MOM AND I ALWAYS TOOK THE TRAIN TO TORONTO. SHE felt it added a touch of fancy, a nice distraction from what was to come. On the train with Mom beside me and Jim on my lap, I'd stare at the pattern on the back of the seat in front of me, wishing I could escape into this world of paisley. Slide down slopes of red petals, dance in swirls of orange, skip along scalloped purple edges. In the world of paisley, everything would smell like butterscotch, and when you licked the red petals, they would taste like strawberry suckers. When you twirled in the swirls, you could make yourself disappear completely.

Before signing into the hospital, we'd stop in at Lime Rickey's in the Eaton Centre for a hot-fudge sundae. The fudge would melt the ice cream into a delicious soup, and for a moment I would forget the dread filling up in my body. My mom always gave me her cherry.

After the sundae, she'd give me a little present. Once, it was white pajamas with pale pink trim to wear in the hospital. They made me feel light and slippery and important. Another time it was pink glittery jelly shoes. They blistered my skin, but when I looked down at them, twinkling along the muted gray hospital floor, I felt their magic.

After Lime Rickey's, we'd head over to University Avenue and walk north toward the blue-and-white sign. "The Hospital for Sick Children." The name always confused me. I never felt sick going into this place; I only felt sick when I left.

I SAT ON THE HOSPITAL BED WITH JIM, WATCHING MY mother pull a sheet over the cot she'd sleep on during my stay. As she slipped her pillow into its case, three men appeared at the door.

"Hello, Katharine!" one of the men said. "How are ya?"

"Fine," I said, feeling suddenly shy. He said my name like we were friends, but none of my real friends called me Katharine.

"Welcome to Sick Kids."

I didn't know yet that this was the nickname for the hospital, so his words hit me more like a warning than a welcoming.

"I'm going to be your surgeon. Remember? I'll be making you the ear."

I did have a vague memory of meeting him the year before, after the plastic ear was removed, but I'd met so many doctors that their faces all blurred together.

He came in closer, pulled out his hand for me to shake. "Put 'er there, kid." I placed a limp hand in his, and he squeezed it, jiggled it, and let go. My hand fell to my side like a dead squid.

"Not much of a grip there, eh?" He laughed, and I noticed the deep half-moon lines carved under his eyes and around his mouth. I noticed the three puffs of hair sticking out from the top of his head. He looked a little like a barefaced clown. He was a small man, but there was something big about him. An energy. Like people should clap after he spoke.

"So, let's take a look at things." The surgeon pulled my hair back and pushed into the side of my head like it was a hard lump of clay.

I looked to my mom. She was leaning against the metal dresser where my nightgowns now rested. Her arms were crossed tightly. She nodded at me as if to say everything would be okay.

It didn't feel okay.

"Stay still, dear," he said as I tried to pull my head away from him. He turned to the other men. "She's already had synthetic implants, but they were rejected."

The men nodded. I mulled over the word *synthetic*.

"So, first we're going to insert a tissue expander here to stretch out the skin for a couple of months." He reached into his front pocket, took out a pen with the cap still on, and made invisible roads on the side of my head. I could tell by how the pen scraped my skin that the cap had been lightly chewed.

"And then"—he brought his pen down to my lower chest—"we'll take cartilage from the rib cage. Here, feel, the ribs are pretty soft." He pulled up my shirt and motioned to his men to come closer. They each took turns pushing into my bottom ribs. I could feel the cartilage bend to their touch. I could feel the bone curve into my guts. I wanted to curl into myself, turn into a ball like an armadillo.

My surgeon continued to talk, explaining his plans to his men. They nodded and wrote things down on clipboards.

When he was finished, he patted me softly on the back. "We'll see you tomorrow, kid."

The group of men straightened and walked to the door.

My surgeon turned back. "Oh, and just so you know, my patients call me Uncle Louie." He winked and slipped the pen back into his front pocket, its roads still blazing on my skin.

I WOULD GET TO KNOW THIS MAN WELL IN THE YEARS THAT followed. He was a man who could twist my stomach into knots just by entering the room, a man who tried his best to relate to me through bad jokes and kid jargon. A man who, no matter his efforts to be friendly, was still the man who sliced into me while I was asleep.

WITH UNCLE LOUIE, I BECAME TWO BODIES: THE ONE I EXPErienced and the one he measured.

The one I experienced flushed pink when excited, nerve endings shining. It knew the interplay of skin and sun. The warmth, the tingling.

The one he measured was skin that split easily. It was trails of sewn flesh mapping the damaged bits.

The one I experienced made puppets of its fingers, mouths of its knees. It hummed songs it didn't quite know the lyrics to, feeling the purr in its face.

The one he measured was arms and legs flailing and stiffening. The thump on the examination table. The slack of the upper lip in the operating room.

THE MAN AT MY BACK PUSHED THE STRETCHER FORWARD.
The wheels rumbled and screeched below me. I held Jim tight, making his head bulge with stuffing. My mom had explained that Uncle Louie and his team were going to be inserting a balloon in the side of my head. She'd tried to make it sound fun, like a birthday balloon. I knew it wasn't going to be like that.

We reached the elevator, and my mom started talking to the man at my back. Her voice was squeaky and weird. She asked him if he'd had a busy day, and he said yes, lots of surgeries today. The elevator dropped down and down and it felt like there was a marble rolling around in my guts. The elevator doors opened. The hallway was white but dark, like the shadows in the corners were bleeding. I looked to my mother. I wanted to tell her not to leave, even though I knew she couldn't stay with me for much longer. The man kept pushing me forward, and the doors at the end of the hall came at me too fast. After these doors would be other doors. Then the green people, the masks, the needles, the rubber hands, the machines, the wires.

I reached for my mom's hand and gripped it tight, hoping to squeeze out a little more time. The stretcher stopped. *All right then, Bunny, we'll see you very soon*, she said. Her smile was too big, like it'd been drawn with a fat red crayon. I wanted to say no. I wanted to say it very loud. But I didn't. *A snap*, she said, *it'll be over in a snap.* She always said that. I didn't let go of her hand. *Okay, Bunny.* She pulled her hand away. *Okay*, she said one more time and then turned around. I watched the back of her as she walked to the elevator. She moved away quickly and didn't turn around.

"IT WASN'T LIKE THAT," MY MOTHER SAYS WHEN, YEARS after the surgeries end, I ask her why she was always cheery when walking me down to the operating room. "I was so scared and I didn't want to make you scared. I wanted to make it feel okay."

"It didn't feel okay," I say.

"I know. It didn't feel okay for me either. After I'd leave you, I'd run to the closest bathroom, turn on the tap, and cry."

"I didn't know that."

I picture her in a gray-toned bathroom, collapsing over a sink. I picture her tears mingling with the tap water—salt with fresh—and I realize something: These surgeries happened to her, too.

ONE TIME, BETWEEN OPERATIONS, MY FRIEND LINDSEY AND I played "hospital." Since I was older, I got to be the doctor and she had to be the patient. The game was fun at first. I pretended to listen to her heart and she mimicked a heartbeat that sounded more like a frog. We laughed. I took her temperature with a popsicle stick.

"Yes, you are very sick," I said. "You need an operation."

I put her on the operating table (a towel on my bedroom floor). She giggled.

"No moving!" I said.

She started swaying her arms.

"I mean it, you have to be very still or I'll have to give you a needle."

This made her laugh, and suddenly I was furious. Didn't she know she wasn't supposed to move in an operating room? Operating rooms were not funny. I stormed out of my bedroom and down the hall to the kitchen, where I found a syringe in the drawer where we kept old keys and elastics. It wasn't a syringe with a needle on the end; it was a hollow tube that was given to me once to measure out liquid medicine after an operation. I held it under the tap and pulled back the plunger to fill the syringe with water. It looked enough like a real needle that I knew Lindsey wouldn't know the difference. It was important that she understood the seriousness of an operation.

I marched back to my room. Lindsey was still on the operating table, but had grabbed two My Little Ponies that were now galloping across her stomach.

I showed her the syringe. "I'm sorry, Lindsey, I'm going to have to give you a needle to put you to sleep."

As I'd predicted, she didn't notice that there was no needle attached. Her smile dropped and her bottom lip quivered. "No, I don't want a needle!"

"I'm sorry, you have to have it. You don't have a choice."

She scrambled to stand up. I could see the tears rolling down her pink cheeks.

"Don't be a baby!" I yelled as she bolted out of my bedroom, down the hall, and out the front door.

The wound was closed with vertical mattress prolene sutures and topical antibiotic applied. The patient was then resuscitated and taken to the recovery room.

It's the word *resuscitated* that stops my breath years later when I read my medical records. General anesthetization, I later learn, is akin to a medically induced coma. Patients are unable to breathe on their own or regulate their own heart rate. Brain activity drops significantly.

It never felt like sleep, more like a blinking in and out of existence. No dreams, just blank.

"Every time you have a general anesthetic," writes Kate Cole-Adams in her book *Anesthesia*, "you take a trip towards death and back."

In the operating room, I was nowhere. Only a body remained. There are stories it knows that I never will.

THE LESION

Within days of the dermatologist visit, the sleep jerks begin. Five to ten times a night, I bolt awake just as I drift to sleep, as though an electric current is rushing through me. I feel betrayed by my body. It's increasingly unsafe. Again.

The internet tells me I'm experiencing hypnic jerks. The internet offers me a quiz: Do I eat a lot of salt? Sugar? Carbohydrates? Caffeine? Do I exercise excessively? Do I exercise before bed? The answer is no to all.

The internet does not ask me why my body, with its old maps aflame, might be afraid to go to sleep.

Eight

ONCE, A COUPLE OF YEARS BEFORE I STARTED THE TO-
ronto surgeries, I turned into a cat. I was on my driveway, and Luke
and the neighborhood boys formed a circle around me. I squeezed
my eyes shut. I had to keep them closed, the boys said, or else it
wouldn't work.

I felt hands on my shoulders, twisting my body.

"Spin around," Luke said.

I began to spin, my feet stepping over themselves, my arms slicing
the air. Just when I started to feel dizzy, there were hands stopping
me, keeping me upright.

And then.

"Wow, look at her tail," I heard.

I felt it, down by my buttocks, a tingling like something was growing.

"And her fur, it's so long."

I felt hairs sprout from my arms and legs, and suddenly I was
very warm.

Other voices:

"Look at her gray head."

"And her pointy ears."

"And her purple whiskers!"

I kept still, letting their words shape my body. Letting their words
tell me what I was.

"What color are my paws?" I asked.

"Cats can't talk, Kate."

My throat tightened.

"You have white paws."

"And a black stripe on your tail."

I squeaked with glee. Then I started to worry. What if they couldn't turn me back into a person? What if I was stuck like this? What if I could never talk again? Would my mom recognize me? Would I have to eat in the basement? Would I have to start pooing in a box?

My cat chin shook.

"Okay," I heard Luke say. "We're turning you back into a human now."

Hands on me again. I spun. This time the dizziness was stronger and I felt like I was falling. The hands kept me straight.

"Okay, you can open your eyes now."

I flapped my eyes open and squinted. The sun was much brighter than I remembered.

"There. See, Kate?" Luke said. "You're human again."

AFTER THE TISSUE EXPANDER WAS INSERTED WHEN I WAS seven, it needed to be inflated. Three times a week, I would go to the hospital in Kingston. My old surgeon, Dr. Winston, would inject saline into the side of my head. He seemed happy to see me—happier than before, when it had been his job to build my ear.

The tissue expander grew with me lying on the examination table, shoulders curled up, wincing, waiting for the sting of the needle, the chill of foreign fluid. It grew to stretch the skin for the future insertion of rib cartilage. Pressure against my head like a firm hand, making the whole right side of my face puffy. Shiny bluish skin stretching beyond its limits. I could hear my blood squeaking through the ever-shrinking channels of veins.

I WAS PICNICKING ON A QUILTED BLANKET WITH MY PARents in the park when I saw a friend running toward us.

"Hi, Kate!" She was wearing a pastel candy bracelet that rattled with her waving hand.

"Hi!" I said. "Mom, Dad, this is my friend Courtney. From Grasshopper Camp last summer."

"Hi, Courtney, nice to meet you," my mom said.

"Yes, nice to meet you." She turned to me. "Kate, my parents are over there." She pointed across the playground to a man and a woman sitting on a green blanket. "I told them about you. About your ear. I want to show it to them. Can you come over?"

"Okay." I stood up and started walking with Courtney across the playground.

"Wait!" I heard my dad call. "Come here a sec."

I pretended to not hear and followed Courtney through the grass toward her parents. Her mom looked up as we approached.

"Mom, Dad, this is the girl I was telling you about from camp," Courtney said, "the girl born without an ear. Show them, Kate." She bit off a candy from her bracelet and looked at me expectantly.

I pulled back my hair and twisted my neck to show them the lump of ear half filled with saline.

"Oh . . . okay." Her mom's face was pink.

Her dad looked at the ground.

After several seconds of silence, I released my hair back to my shoulders. I could feel my face flush warm. "Well . . . I'm going to go back and sit with my mom and dad."

Her mom nodded. Her dad took his gaze off the ground. "It was nice to meet you," he mumbled.

"No, wait!" Courtney wrapped her hand around my wrist. "Tell them about the plastic ear that wouldn't stay stuck to your head!"

"Courtney! That's enough, let her go." Her mom's face got redder.

"But Mom, you wanted to know!"

I didn't know whether to tell her mother about the plastic ear or leave. Courtney's mom's face was so red that I thought it might melt off her neck.

Her dad stood up, released Courtney's hand from me, and said, "Your parents are probably wondering where you are."

My mom and dad were looking at me with stern faces. I walked back over to them, not saying goodbye to Courtney or her parents.

When I got back to the quilt, my parents didn't say anything.

"What?" I asked.

My mom glanced at my dad before talking. "You don't have to do that if you don't want to."

"Do what?"

"You know, show people your ear just because they want to see it."

"Oh."

I didn't know that was an option.

MY DAD AND I LIKED TO GO TO GARAGE SALES AFTER karate class in the summer. We'd scour through other people's junk, looking for our own gems. Once I bought a pair of roller skates from the 1920s for fifty cents. They were metal and clipped onto my shoes. I never wore them outside because the wheels were rickety and uneven.

One Saturday, at a garage sale near my school, I found silver dangly earrings on a faded wooden table. Clip-ons, twenty-five cents. I'd never had my own pair of earrings before. After I handed over my quarter, I clipped one of the earrings to my left ear. "Beautiful!" my dad said. I didn't take it off for two days, only removing it when it was time to wash my hair.

Soon after, I went to a secondhand clothing store and got my mom to buy me another pair: green circles inside of bigger green circles. I asked for clip-on earrings for my birthday, for Christmas. I started wearing them to school. I started wearing them everywhere. The earrings I liked were big and loud. With a big and loud earring on my left side and nothing on my right, the asymmetry was even more noticeable to everyone but me.

Wednesday, January 21, 1987

My favrite things are earings. I have more then thirtyfive pairs of earings. The resin I like earings is because they look nice on my ear. I liked earings since I was in grade two.

School diary entry

NOT LONG AFTER THE TISSUE EXPANDER WAS INSERTED, I was chosen to sing a Cabbage Patch Kid song in my school's talent show. On the night of the big performance, I pulled my hair up in a bun and fastened a pink plastic-jeweled clip-on to my left ear.

"Not tonight, Bunny," Mom said when she saw me.

"Why?"

"Just not tonight."

I pouted. She always let me wear my hair however I wanted. She always let me wear my earrings. I continued to pout as she gently pulled my hair into two low ponytails tied with fat, red ribbons that covered the growing bulge on the right side of my head. I slipped off the earring.

On the stage, with the lights burning my cheeks and the dark cave of audience in front of me, the ribbons rustled against my chin every time I opened my mouth. I felt I had to sing as loud as I could.

"WE WERE FINE WITH LETTING YOU DO YOUR OWN HAIR AS a kid. You liked to wear it up with a bright, dangly earring," my mother says years later. She tells me one of my schoolteachers brought it up to her and my father once, asking if maybe they should do something about it. "We just laughed and shrugged."

I chuckle, pleased with my childhood audaciousness.

"The talent show felt different. You were so cute—you liked to sing with a country twang. You were really excited to be part of the talent show. You practiced in your room every night. We didn't want the ear to be the focus. We wanted you and your adorable little voice to be the focus."

"I get it," I say. "You were trying to protect me."

"Yes," my mother replies. "Trying."

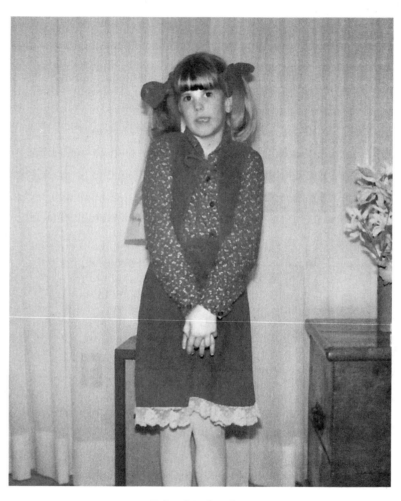

Before the talent show

ONCE A WEEK, I SAT IN THE BASEMENT OF A CHURCH AND listened to my Sunday school teacher talk about Jesus. We learned about how he saved people from sickness and turned water into alcohol. We learned how he was killed, then came back to life. We read passages and answered questions about them. Who came to see Jesus when he was born? How many disciples did he have? We never got to the good stuff, like whether Jesus was ever sad about being different or what it felt like to be healed by him. Or whether Jesus looked rotted out and scary after being dead for three days.

One Sunday, our teacher announced we had a very special person in our class.

"Is it Jesus?" one of us asked.

A good guess.

"No. Someone sitting right here with us today."

This perked us up. We all looked each other over carefully, trying to find any signs of specialness.

"It's Kate," she said brightly. "She's had to spend part of her life in a hospital having operations. She just had one a couple of weeks ago, isn't that right?" She turned to me.

"Yeah." I felt all of the eyes on me.

"How many operations has it been? Tell the kids. How many?"

"I don't know. Seven, I think."

"Seven! Wow. What a brave girl!" She turned to the rest of the class. "Sometimes people are born different, like Kate. But God loves her, too."

The *but* stuck into me. Like God loving me could be an exception. My classmates stared, their mouths tight. I couldn't figure out if they thought I was special or gross.

ON ONE OF MY VISITS TO THE HOSPITAL TO HAVE SALINE injected into the tissue expander, there was another man in the examination room with Dr. Winston.

"He's a resident," Dr. Winston explained. "A student doctor. He's here to learn from you."

Learn from me? I thought. *What can I possibly teach a grown-up?*

The resident nodded at my mom and me.

"Katharine is a little dancer in training," Dr. Winston said, a warm smile on his face.

"We had our recital a couple of months ago and I got to wear a tutu!" I told the resident.

"Oh, how nice," he replied.

"Want to see my pirouette?"

"Maybe you shouldn't be pirouetting just yet," Dr. Winston said. "How are you feeling?"

"I still have a balloon in my head."

"That's right, and we're going to keep filling it up! Let's take a look."

I hopped onto the examination table. Dr. Winston turned to his resident and started to speak about me in a language I'd never heard before—or at least never paid attention to.

His words for me reminded me of the scientific names for bugs on nature shows. Long, sharp, ugly.

The resident scribbled in his notepad as Dr. Winston pointed out all of the things wrong with me.

I felt ugly and dirty, like a slug in a girl suit.

He told the resident about the operations I'd had, all the failures.

The resident kept scribbling, and I thought about how much work it was taking to make me normal.

On our way back to the car after the appointment, I fell a step behind my mom and focused on the stones on the sidewalk.

"What's up, Bunny?" my mom asked.

I shrugged.

"Everything okay?" she asked.

"I don't know." I wasn't sure how to express the feelings swirling inside of me. Instead, I asked a question. "Why didn't you and dad just put me to sleep when I was born?"

After that, my mom didn't allow residents to accompany Dr. Winston on our visits.

THE LESION

I'm back in my doctor's office, just days after I saw her for the swelling and ear blockage. As I sit across from her, I press my toes into the floor to try to feel the ground through my shoes.

I'm corroding, yet every part of me is screaming, twitching, jerking. Like the maggots from my dream are both puppeting and feeding off me. I don't want to breathe. The fact that my lungs continue to inhale is a terrible betrayal.

I can't live in this body anymore.

After I say it, after my doctor's face tenses and recovers, she pulls out a checklist from her desk drawer and starts asking questions.

How long have I felt this way? Do I have a plan to die?

A while. No plan to die, I just don't want to live in this body anymore.

She tells me I have Depression. It will be years before I get a proper diagnosis of Post-Traumatic Stress Disorder, years before my childhood medical intervention will be implicated in my adult functioning. My doctor tells me not to worry; she knows how to make me feel better: Selective Serotonin Reuptake Inhibitors.

The hair was markedly bound down in the area of the incision, which was now covered with raw granulation tissue. The possibility of infection was considered.

The shame grew for years, infecting my blood, breeding fungus on bone. Pushing up against my skin from the inside. Pressurizing. My whole body engorged with it.

Something is wrong with you, it said. *Something is very, very wrong.*

IT WAS A FEW DROPS ON THE PILLOW WHEN I WOKE UP. TINY, like tears on a tissue. When I showed my mom the pillow, she looked at the sinking bulge on my head and her face turned white.

"Oh God, the tissue expander. It burst," she said through a small knot in her face.

I got to stay home from school and sat in the living room, cat on my lap, reading *The Lion, The Witch and the Wardrobe* as fluid quietly leaked from my head. I'd read the book twice before and wished desperately that my cat, Ed, would one day open his mouth and say something smart like Aslan the lion. Mostly he just slept and licked himself.

I snuck to the kitchen and grabbed a Fruit Roll-Up and a chocolate Wagon Wheel. When my mom caught me eating, her mouth knot returned.

"Don't eat anything!" she said. "You might need to have an operation tonight."

The chocolate wheel spun in my stomach.

"What?" I whispered.

A couple of hours later, I was in the backseat of the car, on my way up to Toronto for an emergency operation to remove the tissue expander. I stared out the window, arms folded tight in front of me, trying to get myself used to the idea of having another operation to undo the last one. To start over again.

A RAGING INFECTION. WET SKIN PEELING AWAY IN SLIVERS, revealing the curve of the tissue expander, quietly leaking. My body, once more, pushing the foreign object out.

Emergency surgery. Shuttled into light, my gown pulled to waist, bare chest stuttering shallow exhales. Hands marked blue roads on my head and neck. I squirmed and hands pressed down; voices said, *Be a brave girl.* A moan hot in my throat, a mask clamped over my mouth. *This will help you relax.* I'd never had the mask before. My breath now chemical like burning tires. I shook my head; I wasn't ready. *Time to say good night, Katharine.* But I held onto the room, the lights, the metals, as the edges of me numbed. *Good night.* I held on because I knew when my body stilled, it would no longer be mine. *Good, dear.* The words above me like half-dead balloons. *Good.* The room brushstroked, the body turned cotton.

Nine

FOR MANY YEARS OF MY CHILDHOOD, MY FAMILY RENTED A cottage on Bobs Lake, about an hour east of Kingston. On the way up to the cottage, my dad would stop at a little wooden store with a red roof and pick up a Styrofoam bowl of dirt and worms.

One summer, he made me a fishing pole out of a bamboo rod. We sat on the big rock at the cottage that jutted into the lake, and he showed me how to thread a worm on a hook. I watched the worm wriggle and swell in the places it was pierced and I felt a burning in the places where I had been pierced. Vomit rose in my stomach.

"Don't worry," my dad said when he noticed my upset face, "worms don't feel pain."

How does he know? I thought. Maybe their mouths were just too small to scream.

"I don't want a worm," I said, shaking my head fiercely.

"How do you plan on catching a fish?"

I didn't plan on catching a fish. The thought of pulling out a thrashing animal heaving for oxygen horrified me just as much as threading the live body of a worm with a metal hook.

Instead, I sat on the big rock and dipped the naked string of my bamboo fishing rod into the lake. The string floated on the surface, not heavy enough to submerge. That didn't bother me. It was the water that interested me most. How the sun twinkled on the waves.

My dad let me save a few worms from the Styrofoam bowl. I dug a hole and placed them in the dirt beside the porch.

Luke had heard that if you cut a worm in half, it became two bodies, two distinct worms. I wouldn't let him try with one of my worms, afraid both of the idea that it wasn't true and the worm would die, and of the idea that it *was* true and the worm would have to live the rest of its life split in half.

The following summer, my dad gave up fishing with worms. Still, every year on our way up to the cottage, he would stop in at the little wooden store with the red roof and buy a tub of them. These ones were for me to save. As soon as we reached the cottage, I'd dig a hole in the yard, dump them into the dirt, and watch them burrow to safety.

Fishing wormless

AT A ROUTINE CHECKUP WITH MY PEDIATRICIAN WHEN I was eight, I ran away.

"Now, let's see what they've done here with the ear," he said, reaching to pull back my hair. He was always curious to see the results of Uncle Louie's latest operation. Uncle Louie had taken out half the balloon in the emergency surgery. He'd sealed it off in the hopes of salvaging the rest.

My pediatrician's hands raked along the newly cut skin, pressed into the balloon. I squeezed my shoulders, and my mouth formed a tight ball.

"What's up with you today, Kiddo?"

I didn't know what was up with me. All I knew was his hands were making my bones want to jump out of my body and skitter away. I focused on the poster pinned on the wall in front of me. It was a monkey dangling from a rope. His lips were pulled back in a big smile, revealing too many teeth. His eyes screamed. The caption read: "Hang in there!" I held my breath.

My pediatrician's hands retracted. "If you don't breathe, Katharine, you'll die."

He cocked an eyebrow. He was trying to be funny, but I found it oddly comforting to think I could die right here on the table, numbed to his touch.

A rough finger returned to raw skin. My insides curdled. Then my feet were on solid ground, sandals flapping, floor whirring beneath. I reached the door before realizing what I was doing. I was running away. Past the door, past the nurses' station, past the waiting room. Then the stairs, my legs like shovels digging down.

WITHIN TWO WEEKS, THE AREA AROUND THE REMAINING tissue expander lit up into a blaze of red and I was back in the car on my way to Toronto.

When Uncle Louie examined the ear, he sighed and turned to my mom.

"Have you been cleaning this?" he asked.

"Yes, of course," she said.

"With soap and a cloth?"

"Yes."

"And then the Polysporin?"

"Yes."

"Twice a day?"

"Yes, twice a day." Mom's words were becoming increasingly spiky.

"Well, it doesn't look clean, and it's now infected again." Uncle Louie shook his head. "Did the nurse go over the cleaning procedures with you before you left the hospital last month?"

"I've been cleaning that ear for four years," my mother said.

"It doesn't look clean. And now we're going to have to take the rest of the tissue expander out. Maybe someone needs to sit you down and teach you how to properly clean an incision." Uncle Louie folded his hands in front of him. He looked at me and sighed. I looked away.

Mom tucked her lips inside her mouth and clamped on them with her teeth. For a moment, the room was silent. Then she leaned in to him, her voice just over a whisper. "Maybe someone needs to sit you down and teach you how to talk to people."

"HE WAS AN OLD-SCHOOL DOCTOR," MY MOTHER SAYS NOW of Uncle Louie. "He liked to be in charge. He had a bit of a . . ."

"God complex?"

"You could say that. There was definitely a big ego there. But, you know, when we first met him in his office, he had a carving of praying hands on his desk. That made me feel good about him. Like he had faith in more than himself."

"Do you remember the time you fought with him?" I ask.

"When he accused me of not cleaning the ear? Ugh, I was so frustrated. There we were, our third trip up to Toronto in a month, that tissue expander causing all sorts of chaos, and he needed to blame someone. I felt like a bad mother. Of course I was cleaning the ear. Did he not think I cared? Did he not think I would do anything I could to keep the ear from infection? I wanted to slap him."

"I remember."

"He wasn't a perfect man, for sure. Dad and I believed in him, though. He was passionate. He wanted to make you the greatest ear he could. He wanted the best for you."

"What he thought was the best for me."

THE DAY BEFORE THE REMAINDER OF THE TISSUE EXPANDER was removed, a nurse came to take my blood. She told me it'd just be a "little pinprick." I knew she was lying and I was very tired of the constant poking over the past month. Grudgingly, I offered my hand. She gripped my finger and counted down.

Three, two . . .

I yanked my finger away before the nurse could get to one. I didn't mean to do it, I really didn't. My finger just reacted. Before I could say sorry, I saw the blood. She got me. Then I looked at my finger: no blood.

Uh-oh.

The nurse squinted in pain as she grabbed her own finger.

THAT NIGHT AROUND MIDNIGHT—ME TUCKED INTO MY HOS-
pital bed, my mom asleep on the cot beside me—the lights suddenly flicked on. A man I'd never seen before entered the room. He
approached my bed.

"Hello, Katharine," he said. "I work with Uncle Louie." He was there
to examine me before my operation the next morning.

"Just need you to pull off your nightie," he said.

"No." It wasn't me who said it. My mom stood above me, hand on
my chest. "It's midnight. She's eight years old. She's having surgery in
the morning."

"Ma'am," he said, "she needs her pre-op exam before the surgery."

"Then you should have gotten here earlier."

He stood, dumbstruck, unsure whether to continue to argue or
leave.

"You can do your exam in the morning." Her face was stone.

He stood a little while longer, then turned around and left.

Mom readjusted her pillow on her cot, her curls wild. I smiled into
my blankets. Maybe doctors didn't always get to be the boss.

The patient has a questionable allergy to Lacri-lube.

Drops placed in my eyes to prevent them from drying out during surgery. After, my eyes red-rimmed, puffy, swollen. My vision blurred. The right eye so irritated, I wore an eye patch for almost a week.

Questionable because Lacri-lube is not a known allergen.

Questionable because the only evidence was my eyes.

AFTER THE SURGERY, THE TISSUE EXPANDER NOW FULLY removed, the stretched-out skin sat on my head like puffed pastry. It looked strange, like the skin was melting, but I felt lighter and more myself than I had in a long time. No balloon, no plastic. Just my own squishy skin.

"We'll have to wait a while for things to heal over. There may be enough skin here to continue with grafting the ribs," Uncle Louie said to me with a small grin.

Part of me was relieved. The other part was terrified.

I DIDN'T KNOW TO QUESTION THE SURGERIES. I LEARNED I needed them alongside learning I was a girl named Kate living in a place called Kingston. Needing surgeries was woven into the fabric of me. My knowledge of the missing ear was inextricably linked to my knowledge that it needed to be fixed. I'd never known myself without the fixing part. When the surgeries failed, I blamed my body for failing me, just like it had failed to give me an ear in the first place. It never occurred to me to think my body was protecting me.

I wanted the ear. I hated the surgeries, but I believed they were my salvation from my deviant body. My hope for humanness, for acceptance. For love.

BACK HOME AFTER SURGERY, MY FRIEND ELENA CAME OVER with a large yellow envelope stuffed with homemade cards from my second-grade classmates.

"We spent two whole art classes making these for you!" She beamed as I opened the envelope and spread the cards out in front of me. Folded pages of pictures and words drawn with colored pencils decorated my bed.

There was a card with a carefully drawn portrait of Garfield sitting in a yellow hospital hallway. Words spilled out of his mouth. "Get well soon!" There was a card from Jason, the wheat-haired boy I had a crush on. On the front of the card, a little fish said, "I like you," and a tinier fish behind it said, "I like you, too!" My favorite was a card with a row of bright colorful flowers and a thoughtful message: "I hope you are not bored."

Many of the cards had drawings of me in a hospital, on a bed, thermometers in my mouth, or hot-water bottles on my stomach. In one, there were doctors in hats with crosses surrounding my bed, hugging me with hearts floating out of their heads. In another, I was in an operating room; my eyes were X's, the sun was beaming down, and there was a tree in the corner. Some cards had me dancing with nurses in the hallway; others had me surrounded with big-bowed presents.

It was funny to see what my classmates thought it was like in a hospital—cartoony, colorful, happy.

My favorite card

THAT HALLOWEEN, I DRESSED UP AS A PRINCESS. I WORE A pink shower curtain and carried a wand made of a dish sponge. I wore a plastic mask with a bone-colored face and pink lips upturned into a little smile. I saw through eye slits and breathed through tiny nose holes. The nose holes weren't quite in the same spot as my own nose holes, so I had to suck in hard for air. As the night wore on, the thin space between the mask and my face became hot and stale. The elastic stretched out, making the mask slip to my neck. I had to constantly push it up. Eventually I got tired of the mask and no longer cared about being a fairy princess. I let it slip, and the real me peeked out. The world opened up: houses, sidewalks, streetlamps, stars. The night sky hit my skin like fresh Kool-Aid. I breathed in the new air, and with it, every part of me shimmered.

Halloween princess

I WAS IN MY TWENTIES WHEN BRITNEY SPEARS SHAVED HER head. The images were everywhere: her in a gray hoodie, electric razor in hand, eyes firm on the mirror, wide-open smile. Pain and delight playing on her lips.

The news outlets relished her breakdown. She became a trove of material for late-night hosts.

When a paparazzo asked her why she did it, she said, "Because of you."

When the tattoo artist she visited after her shave asked her why she did it, she said she was tired of "having things plugged into" her head. That she didn't want anyone to touch her. She was sick of having people touching her.

I felt these sentiments deeply. A vicarious catharsis. I studied the look of cool determination on her face as she lifted the razor to her head. Her amused expression of unbecoming. Her head a field of brown pinpricks, unencumbered by the weight of hair, ears naked and shining. A woman pushing against pretty. A woman released.

Years later, she will speak of her reasons for shaving her head. In the days following her shave, however, I didn't think of the excruciating weight of this crisis on her mental health. All I thought was: *Did Britney Spears just start a revolution?*

IN THE MONTHS AFTER THE BALLOON WAS REMOVED, I GOT very used to the puff-pastry ear. With nothing packed beneath the skin, my head was a constant cloud, floaty and soft. One day, Elena and I were on the swings at recess. Our legs kicked the sky as we thrust ourselves forward and back and forward again. We were talking about the ear.

"Would it hurt if I flicked it?" she asked, swinging skyward.

"Nope."

"What if I screamed into it? Would it be loud?"

"Nope."

"What's it smell like?"

"I don't know."

"Can I smell it?"

We hopped off our swings and stood next to each other. She pulled back my hair, leaned in, and sniffed.

"What's it smell like?" I asked.

Elena thought about it, sniffed again, then said, "It smells like your house."

I smiled. I liked that it didn't smell like operation anymore.

Ten

THE NIGHTMARES STARTED WHEN I WAS EIGHT. ALWAYS someone chasing me. One time it was the lady next door who baked me cookies after surgeries. In my dream, her eyes were white and her teeth were gray, glistening with drool. Another time it was a nurse from Sick Kids. She'd grown a big, hairy second head that flopped and thrashed as she ran at me.

When I woke up from these dreams, I'd call out to my parents. I'd hear rustling and the door would open, but it wouldn't be my mom or dad—it would be the monstered people from my dream. They'd lurch forward, their arms and legs droopy, faces creepily wet. I'd wake again, this time for real, and run to my parents' bedroom.

One night before bed, my mom sat me down to talk about the dreams.

"Bunny," she said, "these dreams of people after you, chasing you. Why do you let them do that?"

"'Let them'? What do you mean?"

"Your dreams—they're just your imagination, and you can control your imagination."

"Huh." I'd never considered this before.

"Next time someone's chasing you, stop, turn around, and tell them that they aren't real. Then they'll disappear."

"Really?"

"Yes, really."

That night, it was something new. Not a person, but a giant lobster with big red claws and wiggly antennae poking out of it. It had been living in the basement of Sick Kids for years and had found its way to the wards. It'd already killed two kids and was coming for me now, its spindly legs moving fast and spiderlike below its massive body. I was running away from it, down the shiny gray hallway, when I remembered what my mother had told me. I turned around, looked the lobster in its dead black eyes, and told it to stop.

It kept coming after me.

I said it again and added, "You're not real."

"No," it growled back through a sewer grate of teeth, "*you're* not real."

The lobster crawled on top of me, crushing me into the floor. I couldn't move, couldn't breathe. "No! Stop!" My words weren't helping.

"Stop," I croaked out as its claws tore into my chest and stomach.

THE DAY BEFORE A BIG SPELLING TEST, MY THIRD-GRADE teacher told me I didn't have to worry about studying for it because I'd be away for an operation. I looked at her, shocked. "I'm not having an operation tomorrow."

When I got home and asked my mom about it, she pursed her lips.

"Why did your teacher tell you? I specifically told her not to." She sighed. "Yes, Bunny, we're going up to Toronto tomorrow."

I ran to my room and shut the door. They were going to cut the ribs this time. I paced back and forth, not sure what to feel. Excited to be moving forward on becoming a complete girl? Sad about losing the puff-pastry ear? Scared to have my bones cut?

I didn't know what to do, so I slapped myself hard in the face. The burn made me feel better. My cheek was warm and buzzy. I slapped myself again. Warmth, buzz. I did it again and again until my face felt like it didn't belong to me.

Under general anesthetic, the patient was prepped and draped in supine position with the head slightly tilted to the left.

Supine. A word that conjures an image I'm afraid to see: me on an operating table, body manipulated into a pose it didn't choose. *Supine.* It sounds strangely erotic. Submissive. Unprotected.

MANY YEARS AFTER MY LAST SURGERY, I LEARN THAT UNTIL 2010, it was legal in Canada for medical students to perform pelvic exams on anesthetized patients with vaginas without their consent. It was legal to spread their legs and penetrate them without them knowing.

Even now, in countries where anesthetized pelvic exams without consent have been barred, language around consent can be very vague, with some hospitals asserting that a patient consenting to surgery, or the involvement of medical trainees, is enough to allow these violating acts to occur.

When I learn of this practice, I feel physically ill. Given that I was under the age of eighteen when my surgeries happened, I assume I was not subjected to this nonconsensual penetration. This offers little solace.

What are the values of a system that allows for penetration without consent? That gives precedence to medical education over bodily autonomy?

This is the system my parents entrusted with my body.

THE SURGERY I HAD IN THE THIRD GRADE WAS THE LONGEST surgery yet. The cartilage of three ribs was cut from my lower rib cage, refashioned into the shape of an ear, and slipped under the tissue-expanded skin.

I woke in the hospital bed gulping for air. It felt like a knife stabbing me. A nurse appeared. *Katharine, it's okay. You're okay.* She reached above my head and adjusted the clear mask over my mouth and nose. *Breathe in*, she said. I shook my head. I tried to tell her I couldn't breathe, but a sigh came out of me instead. *I know, sweetie*, she said, *it's going to hurt a bit for a while.* She adjusted a red pouch by my shoulder that attached to a red tube snaking up my neck. *Just a blood drain*, she said when she saw me watching her in horror. *It's okay, sweetie, it's just to suck the blood away from the bone graft.*

Bloodsucker. My chest jerked up. Pain splintered through me. The room shook and smeared.

The wound was closed with a running locked 5-0 Prolene. Patient withstood the procedure well.

False.

THE LESION

In a writing group, I revisit the moment of waking up from the rib surgery. Only the moment, not the reason for the surgery. After I read the piece aloud, one of the women in the room cries and I immediately feel guilty. Perhaps I've overwritten it in a way that makes it sound more horrific than it was? Perhaps she assumes something much worse than what actually happened?

I turn to her. "Sorry, it wasn't—" I try to add *that bad*. But seeing this reaction in another person, crying so openly, I can't get the words out. If I speak, I'm going to cry, too, despite the antidepressant I've just started taking that I assume should shield me from such emotions.

Another woman in the group uses the word *trauma* to describe my experience.

I shake my head and find my words again. "No, it wasn't like that."

When I think of trauma, I think of sexual abuse, war, death, a horrible accident. My surgeries took place in a controlled clinical environment with professionals. How could that be trauma?

Back in my apartment, I pull out my computer and Google "trauma & surgery." Most of what comes up relates to trauma medicine from the emergency room perspective. Eventually I find an article: "When Treatment Becomes Trauma: Defining, Preventing, and Transforming Medical Trauma."

"Medical trauma," authors Michelle Flaum Hall and Scott E. Hall say, "is a disenfranchised trauma—we are socialized to cope with whatever we experience in the medical setting without much thought given to the psychological impacts of treatment and of the medical environment."

A few years later, Flaum Hall and Hall will write one of the only books to date that deals specifically with medical trauma from the patient's perspective: *Managing the Psychological Impact of Medical Trauma*. It will discuss the deep disruption medical intervention can have on the body and mind, even when the intervention is not life-threatening: "Regardless of how noble the intentions, medical trauma exacts a toll that is not easily undone."

The authors will compare medical intervention to certain types of torture. While recognizing that intent and purpose between these two things are very different, they'll contend that both can have similar effects on the body and mind: pain; breach of boundaries; denial of privacy; humiliation; discordant sights, sounds, and odors; strange machines; intrusive touching; altered states; confusion; a body exposed, restricted, constricted, held down, rendered unconscious.

The authors will write about how patients who feel powerless—those who don't have a firm understanding of what is happening to their bodies, those who are not part of the decision-making process—are particularly vulnerable to the effects of medical trauma. For example, children.

They will discuss how trauma experiences can lodge themselves in the muscles and bones, slowly inflaming the body for years. How a routine medical event can trigger body memory, causing the original trauma to reignite and take over. For example, a skin biopsy.

AFTER THE RIB SURGERY, I VERY QUICKLY HAD TO LEARN TO throw up in a new way. I learned to not let my stomach heave or my chest lurch. I learned to lean forward, to slowly rock back and forth to move vomit into my throat. I learned to open my mouth and pant shallow breaths, like a dog. I learned to hold my body stiff as the waves of sick built and crashed and finally made their way out of me. I learned not to cry while doing this, as the trembling of tears only made it hurt more.

That first night after the surgery, I crackled with a pain so present it erased everything around me. I was not a girl with a name; I was not a girl with a past or a future. I was just a body trying to survive.

My mom rubbed my back as I rocked and vomited. Her hand was shaky. I think she might have been crying, but I was too sick to look back at her to check.

"IF YOU HAD A CHILD THAT YOU LOVED SO MUCH, BUT HAD this little physical problem—and you knew she would be judged based on this—would you not try to do something about it? Would you not want the world to love her as much as you do?"

My mother says this to me many years later, when I ask her why we went through with the surgeries. Then she says:

"It didn't feel like a choice. Dad and I were so scared, then Dr. Winston came into my hospital room and said, 'Here's what we're going to do.' And it felt like such a relief, like we could control this. Give you a normal, happy life. It was only supposed to be three surgeries. Just three. It was supposed to be over and done with while you were still very young. Before you entered school, before anyone would notice."

Her voice is quivering slightly, as though she's riddled with guilt. When I was born, she was the first one implicated. When the incisions became infected, when the tissue expander was rejected, she was questioned and lectured.

My father, always a loving and supportive man, worked full-time and couldn't attend the numerous medical appointments and procedures. That responsibility was my mother's. At least that's how I saw it.

My mother came with me to every surgery and slept on a cot beside my hospital bed every night. She took unpaid leave from her work as a social worker, held me when I cried or when I was scared. She worried constantly, cried privately, cleaned the wounds. She did all of this in an attempt to make the surgeries okay. As my mother, she was constantly and implicitly blamed for my body.

Would she have been seen as irresponsible if she and my father had said no to surgical intervention?

A story I once wrote on a box

"LIKE THIS." THE NURSE SAT AT THE EDGE OF MY BED AND took a deep breath. I watched her inflate, then deflate in a series of short coughs. "Now you try."

I did nothing. These exercises, meant to prevent future lung and rib problems, felt impossible. A chunk of me had just been transplanted to the side of my head, and all I wanted to do was lie still and watch *Facts of Life* reruns.

"Katharine?"

I took a slow, shallow breath, just enough that my ribs didn't expand, and released a smooth, horselike whine.

The nurse grinned. "I think you can do better than that."

I grinned back. I wanted to laugh at the funny sound I made, but I was afraid to shake my cut-up body. I tried again, taking a deeper breath. The pain flushed through me like I was breathing boiling water. The coughs were hammers to my ribs. My world became stars and black holes.

"How was that?" the nurse asked.

I shrugged carefully, not wanting to use words. Words required breath. My voice was buried deep, scared to move up my throat and out of my mouth.

"Don't worry, sweetie," she said. "You'll be feeling like your old self in no time."

I thought of the girl shown to me in Dr. Winston's mirror when I was four. Wasn't the goal to become a new self? Was that still *my* goal?

"It'll get better soon." The nurse placed her hand on mine.

It was hard to believe my body would ever feel okay again.

A COUPLE OF MONTHS LATER, MY THIRD-GRADE CLASS GATH-ered with the other primary grades in the school library to watch a special movie called *Kids Can Say No*. At the end of the video, a man with thick glasses sang a song about how our bodies belonged to nobody but us. As he sang, his arms flew up in the air, inviting kids around him to join in.

The melody was catchy, and it stayed in my head for many months. I imagined shaggy beards and crazy eyes and hairy arms that snatched little kids from bushes. I imagined creepy neighbors, fathers, uncles who wanted to touch kids in basements. Truly horrible experiences that I'd never had.

Not once did I think to extend a connection between saying no and my hospital experiences. I knew I wasn't allowed a say in what happened to my body in the hospital. I knew I wasn't allowed to say no when a man in a white coat or green scrubs put his hands on me to examine me or put me to sleep and cut into my skin. I knew in those moments my body did not belong to me.

As a kid, I was not seen as capable of understanding what was best for my body. The task of the doctors was not to get my consent for what they did to me, but to make me as unafraid and compliant as possible. I wasn't told I'd have trouble breathing after the rib surgery. I wasn't told it would hurt for months to sit up, walk, cry, laugh, breathe.

IN HER BOOK ON ANESTHESIA, KATE COLE-ADAMS WRITES: "Our conscious self can only ever tell us part of the story. The rest, that remnant topography, stays submerged beneath the surface. Sometimes it is left to our bodies to do the talking."

The excavation of rib cartilage is an event my body never gets used to. It holds the experience. It remembers.

To this day, the site where they took the rib cartilage remains sensitive. The skin is numb; a dull ache pulses just below the surface. When the area is touched, even lightly, I feel nauseated and unsettled, like I need to leap out of myself. When I twist a certain way, or take in a sharp breath, it feels like a fist in the rib cage.

This is how my body talks.

The event is in my skin, blood, muscles, and bones, reminding me always of the fragility of the human form. After the rib surgery, I never questioned my own fragility.

I will learn that the cartilage removed in that surgery protects the liver from external force. I will learn that my muscles in the area have hardened. My diaphragm is stiff and unyielding, making it near impossible to take a full breath, to take up space. My body shrinks when it should expand.

THE GLASS CASE AT THE ROYAL ONTARIO MUSEUM WAS smaller than I expected. I circled around it three times trying to will myself to look at the mummy inside.

My mom brought me to the museum for my ninth birthday. We spent the morning looking at dinosaur bones, broken pottery, and brass statues. And now we were here. The Egyptian room.

I'd recently become obsessed with Egypt after I'd found out that two of my great-aunts had lived and worked in Cairo for many years. I read all I could about Egypt in our school library.

It was the mummies that fascinated me most—bodies stripped, hollowed, dried, oiled, wrapped, and preserved so their souls would always recognize their bodies when returning from the underworld. I loved the idea of a soul that could take trips away from the body without the constrictions of form. I loved the idea that a body could be preserved for thousands of years in its original state.

When I finally looked inside the glass case, my breath escaped me all at once. What I saw was an unwrapped face with grayish-beige skin and sunken cheeks. I saw parted lips and wooden teeth. Dark caves for eyes, a collapsed nose. It wasn't the features themselves that sucked the breath so violently out of me, but how they came together into an expression of horror. She looked like she was heaving for air. Or trying to scream. Maybe she felt stuck in this glass display, far from the tomb made especially for her. Maybe she was afraid her face, meant to be preserved, would rot now that it was naked to the modern air. Maybe she was scared her soul wouldn't find her in this glass case and would instead wander in the Egyptian sands, forever searching for the body it once owned.

Eleven

ONE NIGHT, I HAD A DREAM I WAS FOLLOWING ELENA through a large plastic tube. She wanted to show me a place she knew. We were excited and running, our heads bent low, our arms stretched out, the edges of us skimming the top and sides of the tube.

Elena started to run faster and then turned back. "Come on, Kate," she said, her voice bouncing echoes down the tube. "Hurry." Ahead, the tube weaved to the right and Elena disappeared. "Come on!"

The tube was getting smaller. I crouched lower and lower until I was on my hands and knees. I crawled, then slithered, to catch up to my friend. The tube shrank more.

"Faster," she said.

I could no longer move.

Elena laughed from a distance. She had reached The Place.

"Kaaaaaate," she called. "We're all here!" I could hear the faint laughter of other kids in The Place.

"I can't," I shouted, my voice garbled as though underwater. "I'm stuck."

THE SUMMER I WAS TEN, MY DAD TOOK ME TO THE PARKING lot behind the old psychiatric hospital and I mounted a red bicycle that was a little too small for me. He held onto the seat and pushed me forward.

"Ready?" he asked.

"Yes!"

"Pedal!"

I lifted my feet to the pedals and immediately dropped them back down in panic as the bike wobbled under me.

"That's okay," my dad said. "That's okay, we'll try again."

He repositioned himself behind me and pushed the bike. My feet rose again to the pedals and, just as fast, struck the pavement when I started to wobble.

"Wobbling is okay," he said. "Just keep pedaling and the bike will straighten itself."

We tried again and I promised myself I wouldn't drop my feet down no matter what. But as soon as I felt the wobble, they dropped before I could tell them not to.

After many attempts, the bike went back into the basement. I tried to forget that I was the only kid I knew who hadn't learned how to ride a bike yet. I convinced myself that my cut-out ribs made my torso uneven and thus I would never be able to balance on a bike.

When I saw kids from the neighborhood riding their bikes outside my window, I'd tell myself bike-riding was stupid. And besides, I'd rather be working on my play. I was writing a script about a girl who had to cut her own arm off because it kept punching her in the face.

Once in a while, kids from the neighborhood would show up at my house to invite me for a bike ride. Excuses would drop out of me. I was tired; I didn't feel like it; I had other, more important things to do. They'd nod and kick off on their bikes down my driveway. I would watch them from my front door until they disappeared down the street.

WHEN WE TOLD OUR FIFTH-GRADE TEACHER THAT MOST OF us watched two to three hours of television a night, he said, "That's ridiculous! How do you get anything else done?"

We shrugged.

"What about playing outside?"

We shrugged again.

He paced at the front of the room, his face pinched like he was working out a puzzle. "Tell you what," he finally said. "If you kids can go one week without watching any television, I'll take you on a special field trip as a reward."

No television for a week? It seemed impossible, but we were intrigued about this mystery reward.

Later that week we learned the reward would be a sleepover on the *Brigantine*, an old sailing ship parked in the Kingston harbor. Staying overnight on a real ship!

Our teacher had been in the Royal Canadian Navy and had connections with the Kingston boating community.

The date for the trip was set for the end of the month, and the TV ban would start in just three days.

I told my mom about the sleepover and she said, "Oh, how neat!"

I told her the date of the field trip and her face lost its juice. "Oh."

"What?" I asked.

"I don't think you're going to be able to go to that."

"Why?" I knew why.

When I told my teacher I couldn't go because I'd be having an operation, he said I didn't have to skip a week of television. I did it anyway. To feel like part of the adventure.

SKIN WAS TAKEN FROM MY NECK AND STOMACH, AND pouched to the bottom of the ear to form a lobe. The ear was now a combination of my buttocks, my ribs, my lower belly, and my neck. Materials taken from my own body. Real skin and cartilage molded into something fake.

Uncle Louie told me that a small percentage of the grafted skin and cartilage would get absorbed into the side of my head. "Your body gets hungry."

He grinned with half his mouth, the way he did when he was trying to be funny, and I thought again of the ghost story I'd heard about the woman made to eat herself.

MY MOTHER HAS A SCAR ON HER STOMACH, A VERTICAL line from belly button to groin. The scar marks the place where doctors cut into her to remove the grapefruit-sized tumor on her ovary—the one that may have led to my missing ear. I was three when my mother had the surgery. Mere months from then, I would have my own first surgery when Dr. Winston would graft the skin from my buttocks.

I used to ask my mother to show me her scar and I'd marvel at how it divided her stomach into two equal parts. Like her belly was the world and the scar, its equator. After my earlobe surgery, I had a scar on my belly, too. Mine was horizontal, marking the boundaries between my stomach and my groin. I liked that my mother and I both had scars in the same place. It connected us. It made me feel less lonely.

So many of my surgeries had unfamiliar consequences for my body—foreign objects blowing up my skin, cartilage sawed and displaced. But here was something knowable. Generational. A scar on my stomach—a scar I'd known on my mother's stomach for years. An experience we shared.

AFTER THE EARLOBE OPERATION, I WAS VISITED BY A RESI-dent of Uncle Louie's who noticed a hunch in my back. He had me stand up and ran his fingers along my spine; my whole back tingled to his touch. He detected a small curve, a place where I crunched into myself. He had me walk away and toward him, noticing a stiffness in my movement. He suspected scoliosis. The word, like the other medical terms attached to my body, sounded alien. He told me I needed an X-ray to confirm. I was distraught. I couldn't afford to have anything else wrong with me.

Three days after my operation, my new incisions on my stomach and neck still raw, I stood as straight as I could in a dark room as machines took pictures of my spine.

The X-rays came back negative. No scoliosis. It was just the way I held my body. Without the diagnosis, the curve was quickly forgotten and I was told I simply needed to hold myself up straighter. With more confidence.

THE BILLBOARD READ: "FREE NEW KIDS ON THE BLOCK Poster with Purchase of a Meal Deal!" I pointed it out to my mom on our way home from the pharmacy, and she pulled the car into the Burger King parking lot. The New Kids had done a concert in Toronto the night after my surgery, four blocks away from the hospital. I had closed my eyes and sworn I could feel their presence. One of my classmates had gone to the concert with her mom and bought T-shirts and took blurry photos of the stage. I'd half hoped the New Kids would stop by the hospital after the concert to get their picture taken with sick children, like celebrities sometimes did. I'd imagined Joey McIntyre sashaying into my room, a single rose in his hand, blond curls glistening, his lips making beautiful shapes as he told me to "Please Don't Go Girl."

My mom and I walked up to the front cash register at the Burger King. The woman behind the register asked for my order. Mid-ask, she looked up and stopped. On her face was an expression I saw often after surgery: eyes growing, then shifting; mouth caught in a soft O. It was the bandages around the head. It was the blood that sometimes crept its way to the edges and dried in red-brown blotches in the gauze. It was the slight limp from the incisions on my chest and belly.

It was hard to know what people thought. Maybe that I had cancer? Or had been in a bad accident? Once, a woman at the grocery store put her hands on me and said something about my poor soul and God blessing it.

I made my voice small and fragile for the woman behind the cash register. "What do I need to order to get the Joey McIntyre poster?"

"Um. Any of the combo Meal Deals." She swung her hand back to point to the menu above her. "But. Let's see." She disappeared under the cash register, then popped back up. "You don't need to get a Meal Deal. I have a poster right here for you, sweetie. Two, actually. Do you want two posters?"

I looked to my mom. She was giving me the "don't be greedy" face. I ignored it.

At home, with my two posters of Joey McIntyre tacked on my bedroom wall, I played the *Hangin' Tough* album and he sang to me in my own private concert.

AFTER THE LOBE SURGERY, MY MOM'S FRIEND BOUGHT ME a pair of sterling silver earrings in the shape of teardrops. Real earrings—not the plastic clip-ons I wore on my left side.

"For when your ear is finished," my mother's friend said when she gave them to me.

I pulled them out before I went to bed, pressed one against my left earlobe and the other against the bandage on the right side where the new earlobe was healing. The earrings winked and sparkled.

"Can I get it pierced?" I asked Uncle Louie at my post-op appointment.

"What?"

"Can I get the earlobe pierced?"

"The earlobe? You want to pierce it? Why would you want to do that?"

"Because I want to wear an earring there. Mom's friend gave me these earrings. They're sterling silver and I want to be able to wear them—you know, wear something shiny there."

His eyebrows pinched toward each other.

"But I just made you this beautiful lobe. Piercing it, I don't know about that. And the lobe's attached to your head. There's no room for the backing of an earring."

Blood drained from my face like a flushing toilet.

Uncle Louie saw my disappointment and put his hand on my shoulder: an act of tenderness I hadn't experienced from him before. "It would mean extra surgeries to pull the lobe out. Maybe grafting more skin, and a small bone graft to lift the lobe away from your head." He started to look interested at the prospect of this new challenge.

"Never mind, it's okay," I said. There was no way I was having any more surgeries than were already on the roster.

But for years, I keep the earrings in a heart-shaped box, along with the cheap metal rings and necklaces that come and go as my tastes change. Even after the shine of the silver fades to a dull gray and the black freckles appear, I keep these earrings. Long after I leave home, these earrings stay in the heart-shaped box in my childhood bedroom, waiting for me.

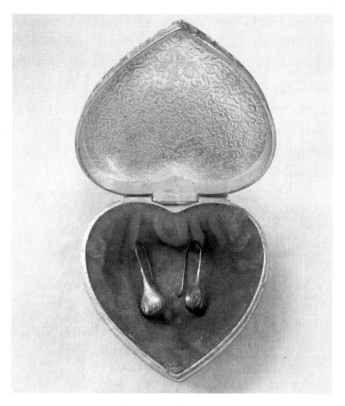

The sterling silver earrings

THE KATE I'M SUPPOSED TO BE HAD BOTH HER EARS PIERCED years ago. She wore studs and dangly earrings, and ones made of white gold and yellow gold and sterling silver. She glittered from both sides of her head. She knew how to ride a bike and she broke through the air at high speed and popped wheelies and rode down to the quarry with the other neighborhood kids. She never missed a birthday party and went to all of the sleepovers and went to New Kids on the Block concerts and slept on boats with her classmates and wasn't that fun when they all took turns steering the ship and snuggled into their bunks that night? She was never scared and she could stand so straight, so confidently, that she was as tall as her dad and she didn't take crap from anyone.

A FEW DAYS AFTER THE VISIT WITH UNCLE LOUIE, I WAS PLAY-ing Monopoly with Elena in my living room. When I returned from the bathroom, she was gone. "Elena?"

I heard faint laughter, but because of my bad hearing, I couldn't figure out where it was coming from. "Elena, where are you?"

"In here," she said without telling me where "here" was.

"Seriously, where are you?" My voice was crispy with anger. "Just tell me where you are!"

I searched the kitchen, the laundry hamper, the closets in my parents' bedroom.

"In here," she sang, sounding farther away.

"This isn't funny!" I was yelling now, marching between the guest room, the bathroom, and my bedroom. What was wrong with me? Why couldn't I just figure out where she was?

More giggling.

Then I saw a hand peeking out from under my bed. My insides swirled as I stepped fully into my room. Without hesitating, I leapt up onto my bed and jumped. Hard. I felt the springs of the mattress sag and retract under my feet. Faster and faster I jumped. A hot rage took over my body. I pretended it was Uncle Louie under there, face flat to the floor. I pretended it was me under there and I was smashing my own body to bits.

"Stop it!" I heard Elena shout from below.

But I couldn't. I was a blur. I was a demon. I had the power this time.

WHEN I RETURNED TO SCHOOL, MY CLASSMATES WERE BUSY writing "Sea Logs," recounting their adventures the previous week on the *Brigantine*. I asked my teacher what he wanted me to do, and he handed me a short stack of lined paper. "I want to hear about your excursion, too," he said.

I'd never thought about my trips to Toronto as excursions. I'd only thought of them as an absence, a missing-out on the adventures other kids were having. It never occurred to me that anyone would want to hear about them.

I took the stack of paper from my teacher's hands and headed back to my desk, dazzled by the idea that I had a story, too. I wrote about the train ride to Toronto. I wrote about being wheeled down to the operating room, about being put to sleep. I wrote about the cafeteria in the basement. I wrote about my favorite nurse, Brenda, who always used very soft hands when cleaning my incisions and stopped whenever I said "Ow" to ask me if I was okay.

I wrote almost six full pages, front and back. I didn't know I'd have so much to say.

My teacher let me read my story in front of the whole class, just as the other students got to read their stories about the *Brigantine*.

One day, at recess, he told one of my classmates that I was very good with words. I glowed with pride. It was the first time a teacher saw me as more than the kid who missed class, the kid who didn't hear well, the kid who needed extra help to catch up. It was the first time I felt special for something beyond my body.

Twelve

IN THE KIDS' GAME PERFECTION, PLAYERS MUST FIT TINY plastic shapes into matching holes on the board. If a player fails to place all the pieces before the time runs out, the board springs up with a loud pop and the shapes explode into the air. After receiving the game for Christmas, I spent many hours a week trying to win, trying to get my pieces to fit—all the while bracing myself for the inevitable explosion.

MY NEW BALLET TEACHER WAS UNLIKE ANY I'D EVER SEEN before. He was a man, for one thing, and had dirty-blond hair swept to the side like the arc of an ocean wave. He had a pointy chin and spoke with a thick Russian accent that made all his words sound important. He never smiled. His name, Yuri, was very fun to say, even though he made us call him Teacher. It took me a while to realize that he looked just like Westley from my favorite movie, *The Princess Bride*.

Our classes with Yuri were giggle-free. He'd count out beats with a clap of his hands and we'd jump and land and jump.

He'd say, "You must know the splits by now."

Most of us couldn't do them, myself included. He'd shake his head at our Canadian training. "No ballet girl in Russia would get away with not knowing the splits."

In the second half of class, he'd come to us, one by one, as we each tried to do the splits. He'd press us down at the shoulders. Our legs would stretch and tremble as they were forced closer to the floor. He'd hold us there until our whole bodies shook. Then he'd move on to the next girl.

Melanie was the best splitter. She was very bendable because she'd taken gymnastics for years before switching to ballet. Yuri loved Melanie. "This," he'd say, "this is what your bodies should do," and he'd place his hands on her back leg (a leg already flush with the floor in a full split) and pull it slightly upward, in a hyperextension. We all admired Melanie and her elastic legs, and the way her face would remain calm and pretty when Yuri pushed down on her.

Janelle was the worst splitter in the class. She kept her legs firm in a tall V, her middle a couple of feet from the ground. Yuri would press her down and her legs wouldn't move. He'd press down and she'd push back up. If he pressed too hard, she'd simply bend her knees and crumble to the floor. Yuri's face would get very red. She'd stay crumbled on the floor, a light smile on her face. Eventually, he started passing her by completely, and she seemed very fine about it. Some of the other girls would roll their eyes at her, but I thought she was one of the bravest girls I'd ever met.

Every six months, we will just keep working away and try to bring her closer to a symmetrical, somewhat ascetic [sic] situation.

Western ascetics of the Middle Ages believed the unruly body to be an instrument of sin. Ascetic rituals—self-flagellation, starvation, solitary confinement, mutilation—were done to discipline the body and purify the soul. To stay the path of righteousness.

AT MY NEXT PRE-OP APPOINTMENT WITH UNCLE LOUIE, HE pointed to my side ponytail and asked why I was wearing my hair like that.

"I don't know, I like it. It's in style."

I'd seen it in *Teen* magazine. A girl with a perfect oval face skipping across the street, head back, laughing, auburn hair swept in a side knot.

"In style? To hide the ear I'm making for you?" He looked hurt. His eyes wide, his mouth pinching at the corners. Under the lines and pockets of his face, I could see the little boy he once was.

"No. I'm sorry. It's just. The style . . . I could've put it on either side, but—"

"But you didn't."

I find out, years later, that Uncle Louie was a wood carver in his spare time and carved ducks and owls and lions and children. In an interview for a small newspaper, conducted decades after I last see him, he compares plastic surgery to his wood carving. Both, he says, require an individual to see the finished project before starting. I imagine him saying this to the interviewer, that amused grin of his stretching across his face. Projects, both: wood carving and body reconstruction.

Was my body his art? What happens to a girl when her surgeon envisions his idealized version of her before he even starts? Who is she to him before he realizes this vision? And what does she become if it is never realized?

AFTER MY FIRST FEW OPERATIONS IN TORONTO, UNCLE Louie decided I should take a pill to settle myself before surgery. I was ten years old and still crying in the operating room, making it difficult for the anesthetist to do his job. I was embarrassed. I wasn't a little girl anymore, and after so many surgeries, I should have been very used to the operating room. Yet, with each passing surgery, I was becoming more and more scared.

After I was told about the pill, my mom said I didn't have to take it if I didn't want to. That if I felt like crying in the operating room, I could.

I didn't want to be a crybaby, though. I wanted the doctors to like me. I wanted to be good.

The first time I took the pill, it didn't work. My dread easily broke through it. My heart still pounded; the gurgle in my belly still threatened to crack open into a cry. But after the whole production of getting me the pill, I pushed the tears back into my guts. I made sure I never cried in the operating room again.

THE LESION

When I next see my doctor to follow up about the Depression, she asks how the antidepressants are working. I don't know what to say because I'm not sure they're helping. If anything, they make me feel canned away from the rest of the world, suspended in syrup; the despair of living in my body now trapped inside Tupperware containers and left to rot, like leftover meat at the back of the fridge. Instead of helping me live in my body, the pills seem to be freezing me out of it.

"Give it more time," my doctor says. "The first few weeks can be wonky. Your neurochemistry is still adjusting. You'll be feeling better soon."

As I cross and uncross my legs in the plastic chair, I think about the pills I used to take before surgery, given to me to mute my body's response to the operating room. A medical intervention to deal with my reaction to medical intervention. And I wonder, with the antidepressants, if we're trying to hush my body instead of listening to it.

IN THE FALL I WAS ELEVEN, I UNDERWENT AN OPERATION TO readjust the earlobe. It was set too low, not quite lining up with the left ear. A surgery to correct the previous one. More skin from my abdomen to add fullness.

I woke up to the familiar sting of cut skin on my belly, the familiar ache of displaced flesh sewn to the ear, the familiar pulse of sick from the anesthetic. Also a new feeling, one I couldn't quite articulate. I woke up with a sense of stuckness. A feeling that I was going nowhere.

THE LESION

The dermatologist calls to give me the biopsy results. There is a precancerous lesion on the earlobe. Sun damage.

"Don't you use sunscreen on the ear?" the dermatologist asks.

"No."

I don't tell him I haven't looked at the ear or touched it in twenty years, that it's been buried under thick drapes of hair. I don't tell him I have no idea how sun could have damaged it, as it has rarely seen direct daylight in decades.

Most of the skin for the earlobe was taken from my lower stomach, an area not equipped to handle sun exposure well. Given the history of the area in question, the dermatologist doesn't feel comfortable removing the lesion. He refers me to a plastic surgeon.

I'D MADE THE MISTAKE ONCE OF LOOKING IN A MIRROR TOO soon after surgery, and what I saw wasn't me but a zombie version of me that scared me for days. It was the eyes that frightened me the most: red, drooped, and glazed gray by the anesthetic. I learned to avoid mirrors and made sure never to glance down to my chest and stomach in those first couple of days following an operation.

After enough time had passed and I felt ready, I'd approach the mirror to discover what the surgeons had done to my body while I'd been unconscious. Sometimes there were bruises on my face or neck, or parts of me were swollen or displaced. Sometimes I was still marked with the slippery brown film of leftover iodine, sitting on my skin like dried urine. The bandages wrapped around my head hid the ear itself, but also crept down my forehead; the gauze and medical tape threatened to obstruct my vision as the days wore on. Blue wiry stitches poked out of me somewhere, usually my stomach, and the raw-steak smell of fresh incisions that marked where I had been split clung to me like an aura.

It took a while for me to recognize myself again. In the days, weeks, and months that followed an operation, I'd see my body adjust to the changes imposed on it. Incisions whitened to scars, lumps flattened, transplanted tissues nestled into their new homes. Pain dulled, itched, then mostly disappeared.

And when my body started to feel familiar again, it would be time for another operation. I'd become alien to myself all over again.

IN GRAD SCHOOL, I LEARNED ABOUT A CONDITION IN WHICH people split off from themselves. For some, it's a complete detachment from their own body, as if it doesn't belong to them; for others, it's one particular body part that splits off. Sometimes it feels like the body is changing shape or growing or shrinking, or that certain parts feel fake or plastic. Sometimes people feel that a particular body part has been transplanted from other areas. Sometimes people feel invisible, not seen or heard by others. Sometimes people are unable to recognize themselves in the mirror.

The medical name of this condition is depersonalization. A literal un-personing of self. This is my relationship to my body after surgery.

I SAT UP IN MY HOSPITAL BED AND MARVELED AT THE FOOD tray propped in front of me. I loved hospital food. I loved how it was expertly placed on the tray in perfect compartments, how the main course was hidden beneath a metal cover, waiting to surprise me, and how no matter what was under there, it always smelled like cream of mushroom soup. I pulled away the cover. Chicken! A cloud of potatoes! Green beans! Laid out so neatly it almost looked plastic.

I sliced my fork into the wet meat, and it cut through smoothly. It wasn't chewy like the chicken I ate at home, but instead turned to mush on my tongue. I picked up the orange cube of Jell-O in the corner of the tray and held it to my eye. The room turned bright and wobbly. Brilliant orange walls, a tangerine floor, a bed made of creamsicles with metal bedposts of cut carrots. In the hallway, I saw an orange man talking to an orange woman.

"And this is 7C, our plastic surgery ward," the orange man said.

"Plastic surgery on children? You mean like elective surgery?" the orange woman asked, looking into the hospital room across from mine.

"Not really. Most of the kids here were born with some sort of physical anomaly. The surgeries are mostly reconstructive in nature."

Just then, the woman looked into my room, and I wondered if she thought I was born with an orange cube on my face.

MY FRIENDS ON WARD 7C WERE, LIKE ME, KIDS UNDER CON-struction. Misshaped bodies gathered and recrafted to match the shapes of other kids. Cut at the edges to fit.

Julia was three. Her hair shaved, a big scar stretched from ear to ear across the top of her head, like a worm fat with rain. She was born with her eyes drooped into her cheeks, so the doctors cracked her skull open to set things right. So she could see like other kids see. She was too sick to leave her bed. Sometimes she'd clutch its metal bars and pull herself up. Sometimes she sang "The Wheels on the Bus" in a squeaky little voice, and we'd come to her room and sing and clap along. She'd giggle and her eyes would roll around like lost marbles, searching for our faces. When we got loud, the nurse would come in. She'd tell us Julia needed to rest. She'd pull Julia's hands from the metal bars and settle her into the bed. Julia would try to close her eyes, and they'd turn into slippery white balls. Her eyelids didn't work yet.

Every time I came back for another operation, the smell of 7C smacked me in the face—bleach, old milk, vomit, the burnt-rubber stench that each of us brought back from the operating room. The smells all mixed together into a soup that flowed through the hallway. After a day or two, the smell seeped into my skin and I became part of the soup.

In the back lobby of the hospital, there were two large display frames of "foreign bodies" that kids had swallowed or jammed inside them-selves. Rotting peanuts, coins the size of Ping-Pong balls, Monopoly pieces, buttons, long rusty nails, and, most unsettling, open safety pins, some as big as a Swiss Army knife. On 7C, we talked a lot about these displays. It was one of the first things we'd ask each other when we met. "Have you seen the Display of Foreign Bodies?" I don't know what fascinated us most, the fact that kids freely chose to stuff these objects inside themselves or the fact that these things didn't kill them. We'd ask each other which of the objects we would choose to have stuck inside us. Most of us said the small buttons or the little dog from the Monopoly set. Some of us, when we needed to feel brave, would say the nails or the big open safety pin.

Tim showed me the thin scar between his thumb and pointer finger as we waited for our pre-op photos outside the photography room. He told me he once had an extra thumb there. Two thumbs on one hand! I wanted to ask if he was mad they cut it off. Couldn't an extra thumb be useful? Instead, I asked if he missed it. He tilted his head, said, "Sometimes." His dad had signed him up for baseball that summer because he could now fit his hand inside a glove. "I'm going to be the next Kelly Gruber," he said. He didn't sound excited. He told me he could still feel it, the cut-off thumb. Like it was a miniature ghost haunting his hand.

There were rules in the photography room. Those of us who had been there before knew the rules well and made sure to prepare the first-timers before they went in. The pointy-faced photographer pretended to be a fun guy, but got annoyed when we didn't do things the way he wanted. In the photography room, we sat on a stool and had to keep our lips relaxed (these weren't the kind of pictures we were supposed to smile in). The photographer bent and twisted us in new ways, and we had to stay very still in whatever shape he put us in. After he fiddled with his camera, he'd say "Boom!," click a button, and the umbrellas behind him would explode with light.

Before our surgeries, we had to put on mint-green hospital gowns because they opened right up and the surgeons didn't have to fuss around to get to our bodies. All of the hospital gowns had "The Property of the Hospital for Sick Children" stamped on the bottom hem in a little triangle. Like anyone would ever steal a hospital gown. None of us liked wearing them because they left us bare at the back, and once we put them on, we knew they weren't coming off for at least a couple of days. Once we slipped our arms through the crisp green sleeves, we, too, became The Property of the Hospital for Sick Children.

Erica taught me how to open the large steel door of the ward fridge so we could swipe popsicles when the nurses weren't looking. We'd creep into the hall, our slippers soft on the speckled floor. We'd look right, we'd look left, and when it was clear, we'd wrap our hands around the fridge handle and pull hard. The purple birthmark splotched on Erica's face would crimp as the door swung open. The cold air would blow over us like the breath of an ice witch. We'd run from the fridge, popsicles numbing our hands, and her blond ringlets would bounce like Slinkys. After my operation, I was too dizzy to move much, so Erica slipped into my room with a cherry popsicle (my favorite!). I slurped it down faster than I should have. It was still cold when I vomited it up a few seconds later.

There was a computer on wheels near the nurses' station, and when we were bored, some of us would drag it into one of our rooms and plug it in. It only played Frogger. In the game, we were frogs that had to cross a busy street and a crocodile-filled river to get home safely. Most of us didn't make it.

I met Becky in the playroom on the sixth floor. She showed me her half-eaten apple, said she ate the missing half in one big bite. I didn't believe her, but I nodded anyway. She had black hair and pink cheeks, and one of her arms was snaked with jagged red skin. "It died," she said of her arm, dimples stamping into her pink cheeks. I asked her who killed it, and she told me it was a mystery. I found out later that her dad did it. Poured hot coffee on her when she took too long to eat her Cheerios. He wasn't allowed on the ward.

I first saw Laura on my way back from the bathroom. I recognized myself in her instantly. Our backs were bent by the same rearrangements of skin and cartilage. Our postures confused, pulling us downward, as if to say sorry. When the nurse told me there was another girl on the ward born without a right ear, I told her I already knew. Laura and I didn't talk about operations or ears. Instead, we crammed ourselves into the movie booth in the playroom and watched cartoons for a quarter. There was room for just the two of us, and our thighs touched and our knees knocked together. It didn't matter that the cartoons were childish. It was enough to be in the booth together. It was enough to know someone else like us existed.

When the pack of surgeons came to the ward, we all got very quiet and hurried to our rooms to wait to be looked at. There was a different smell when the surgeons came—sweat mixed with cologne. Sometimes, a surgeon pressed his hands into a sore spot and a scream would crack open in one of us. The scream would gallop down the hall. When it reached the rest of us in our rooms, lying on our crinkly bedsheets, we'd hold the places we'd been cut, our incisions humming like mouths clamped shut.

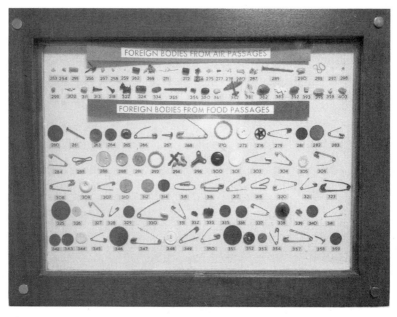

The Display of Foreign Bodies

"ABJECTION," WRITES DEBORAH CASLAV COVINO IN HER BOOK *Amending the Abject Body*, "is created/sustained by the medical field . . . The state of abjection is 'alienation' and the process of abjection (ridding ourselves of the unwanted) is an act of orientation to a welcoming community of clean and proper bodies—medicine portrays the amended body as necessary for communal integration."

For some of us on 7C, our body differences were the result of accidents or events in our lives. The goal in these cases was to erase the story—bring these bodies back to an original, unblemished state. For most of us, however, the variances in our bodies were there at birth. There was no original unblemished state. As sociology professor Heather Laine Talley says in her book, *Saving Face: Disfigurement and the Politics of Appearance*: When the difference is congenital, there is no such original-body reference point. The reference for us was non-variant bodies that were never our own. The reference was imagined, envisioned by the doctors.

"Disfigurement (as a diagnostic category)," Talley writes, "materializes through the medical gaze. Facilitated by medical technologies that frame and focus the physician's optical possession of the patient, the medical gaze abstracts the suffering person from her sociological context and reframes her as a 'case' or 'condition' to be fixed."

All of my surgeries were covered by insurance. All of the surgeries were deemed a medical necessity. Was my missing ear a kind of sickness? In need of intervention in a system that monitors, promotes, and perpetuates the concept of what a body should look like in order to be deemed healthy? What is the line between healing and harming a body?

Thirteen

HE ASKED ME IF MY DAD WAS HOME.

"No," I said, holding the phone awkwardly to my face.

"How old are you?" he asked.

"Eleven."

"Are you starting to get hair in new places?"

"What?"

"Like on your body," he asks.

"A little."

"Where?"

Silence.

"Where, like between your legs?"

"Yes."

"What color is it?"

My mother walked into the kitchen. "Who are you talking to?" The phone went dead.

I told her I was talking to a man; she asked me what he wanted.

"Jesus," she said when I told her. "What did you say to him?"

"Nothing, I just answered his questions."

"Why did you answer his questions?"

"I don't know," I said, prickling with guilt.

I was used to answering questions about my body, used to talking about it like a thing that didn't belong to me, used to seeing through the eyes of the men who asked questions about it, who claimed it with their words and tools.

A SMALL GROUP OF BOYS FROM MY SIXTH-GRADE CLASS gathered in the schoolyard at recess near the bike racks. It was an unusually warm fall day, and I was sitting on the grass nearby with Elena, both of us picking at dandelions as we listened to Richard Marx's "Right Here Waiting" on repeat on our Walkmans. I noticed the boys showing each other raised nicks on their elbows, scabs flaking on their wrists and knees. When I realized they were comparing scars, I abandoned my dandelions, unhooked my Walkman, and stood up.

"Kate, what are you—?" Elena asked without finishing her question.

"I'll be back."

The boys didn't notice me in front of them.

"Hey," I said, louder than I'd meant.

One of the boys, Kyle, looked up from the thigh scab he was showing his friends. "What? What do you want?"

I lifted my blouse to reveal the fresh incision on my abdomen. The area was still red and bumpy, the skin puckered where the wound was scabbing over. Other whitened scars lay above and below like faint chalk lines. Before I could lift my blouse up further to reveal the scar on my rib cage, Kyle raised his hand and said, "Ew, that's disgusting."

Jason—the boy with the wheat hair who once made me a get-well card with a fish that said "I like you," the boy who was now the smartest boy in the class, the boy who made me warm when I looked at him for too long—bent his head and reddened.

I pulled my blouse down and gripped my elbows with my hands, forming a shield across my torso. What had I just done? The boys looked at me like I was a car accident. My face turned to fire as I turned around and walked back to Elena in the grass.

"What was that?" she asked. She'd seen what I'd done and was now looking at me like the boys had.

I didn't have an answer for her. I picked up a dandelion and crushed it into my wrist.

WHAT MADE ME PULL UP MY BLOUSE TO THE BOYS ON THE schoolyard? Why did I need to show them my scars?

I asked myself this question for years. The simple answer is I felt compelled. The more complicated reason is that I needed the boys to see my scars because I needed an outside witness to what was happening when I vanished from school for weeks at a time. I needed them to see that something was done to my body.

My scars were a record of my body changing. Evidence of my original form disappearing into something else. I needed to share my suffering. I needed my peers to see the wound on my skin, to know it was real so that I knew it was real.

ON PICTURE DAY, MY CLASS TRUDGED INTO THE GYMNASIUM for our group photo. The girls slicked on silvery pink lip gloss while the boys challenged each other to shout the word *penis* without being detected by our teacher. I practiced my photo face. A soft smile, no teeth.

Our teacher ordered us by height and marched us down to the benches against the wall where the photographer waited. Because I was tall, I ended up in the far right corner of the back row. Kyle stood to my left. He quickly shuffled himself away from me, creating a gap where a whole other person could have stood.

"You!" the photographer said, pointing to Kyle. "You need to move a bit closer to her." He pointed at me.

Kyle stood still.

"Closer to her," the photographer repeated.

No movement from Kyle.

"What's wrong with him?" the photographer asked my teacher. "Why won't he move?"

"Kyle!" my teacher said. "*Now!*"

Kyle sighed and slunk in closer.

As I waited for the flash, posed and ready with my soft smile, Kyle leaned in.

"It's because you're so gross," he said.

Then, the flash.

Years later, I find the photo in a drawer in my childhood bedroom. It is from eighth grade, not sixth grade like I remembered—two years after I'd shown the boys my scars. In the photo, Kyle is smiling, broad and big-toothed, his hair tousled with gel. I, somehow, have managed to smile, too. Tight-lipped. My own hair is a kinked mess of gold, fried at the ends from a bad perm. I'm wearing a dark sweater that hangs off my thin frame like a lost shadow. My shoulders are bent forward, forming a cave in my chest. There's a nuanced discomfort between Kyle and me—his body angled away from me, and my face tilted slightly down, eyes glazed. But it's nearly undetectable in the photo.

I tried to explain to Kath[a]rine today a few of our objectives and goals and the various stages so as to lessen her dread of the hospital area. She knows that the next stage will be an outset with skin graft taken from the left bikini area.

I don't remember this visit, but I can imagine. My surgeon listing his ambitions for my body. The ways he'll carve into me. My face practicing *brave* as the ghost of the incision yet to come presses firmly on my groin.

ONE DAY, OUR SIXTH-GRADE TEACHER TAPED UP A DIAGRAM that looked like the outline of a goat's head and told us it was the Female Reproductive System. The boys snickered and the girls shrank in their seats. He used the word *vagina* over and over, and each time it hit the air like a clumsy bird.

We learned about fallopian tubes, eggs descending, and bloodshed. He told us periods allowed our bodies to someday grow babies. He said periods were the start of womanhood.

Womanhood. It sounded full-bodied and beautiful.

At recess, Elena told me that periods were called "the curse" because once they started, we would bleed and cramp and ache for seven days every month.

She said "we." *We,* like *us,* like *me, too.* She said "we will," like it was inevitable, like she wasn't at all worried she wouldn't get it.

She looked at the slight frown on my face and said she didn't mean to scare me, but that we should be prepared.

I wasn't scared of getting the period. I was scared of not getting it. Of my body not qualifying for such fancy human things.

FOR OUR YEARBOOK PAGE THAT YEAR, OUR TEACHER ASKED us to draw our bodies below our headshots. Some kids drew their bodies doing something they enjoy—hockey, skateboarding, dancing. Many of the girls drew themselves with breasts, some in bathing suits, others in dresses and skirts, posing with hands on hips.

I drew myself as a fish. A scaly lump with two fins and a triangle tail. The only portrait not human. Slippery, untouchable, unable to sustain life in open air.

THE NIGHT BEFORE MY NEXT OPERATION, UNCLE LOUIE CAME to my hospital room with a very cute young man with thick eyelashes and hair shaved into a buzz cut.

"How are ya, Katharine?" Uncle Louie asked. "Are you a super-model yet?" He liked to ask me that now. I could tell he was trying to compliment me, to make me feel pretty, but all it did was make me uncomfortable.

"No."

"Someday," he said with a wink.

He turned to the very cute man. "One of my residents," he said. "He'll be assisting me in your surgery tomorrow."

The resident nodded. "Hello." His mouth stretched into a dazzling smile.

All I could think about was what it would be like to nuzzle my head into his armpit and leave it there.

"So, let's take a look at you," Uncle Louie said, pulling himself toward me.

He examined the ear like he always did, pushing his fingers into the tender spots. Then he flipped up my nightie and made a gesture for his resident to come closer.

"Here," Uncle Louie said as he pulled my underwear down a notch, "is where we'll take the skin graft tomorrow to contour the ear." He pointed near my groin area.

Hot blood washed the insides of my cheeks. I wanted to hide, dissolve, un-exist. This man with the buzz cut and the thick eyelashes would see my privates when I was asleep. The resident leaned in, his breath warming my stomach. The tang of aftershave bitter in my throat.

As Uncle Louie spoke to his resident, I became two bodies: the one I experienced and the one they measured. The one I experienced had budding breasts, hard as walnuts, a constant ache in the chest. The one I experienced had fuzz sprouting, white phlegm in my underwear, the lightning flush of emotions—sad, happy, humiliated, excited, dejected.

The one they measured was good skin for harvesting.

A PIECE OF RIB CARTILAGE FROM THE EAR WAS EXCISED AND wedged between the top of the ear and my head to add more dimension. Skin taken from my pelvic area was used to smooth the rivets of the ear. Skin blanketing skin, edges becoming soft and round.

When Uncle Louie's resident with the buzz cut and the thick eyelashes came by my hospital room to examine me a couple of days later, I was so embarrassed I couldn't look at him. I stayed frozen as he pressed his fingers into the bandages covering the ear, as he flipped up my nightgown to examine the new incisions he'd made.

"How does it feel?" he asked.

"Fine," I said, my voice barely above a whisper. I imagined myself as a statue, the kind the ancient Greeks used to make. Pure marble, polished and numb.

ONE TIME, IN MY LATE TWENTIES, I WAS ON A CROWDED SUB-way and felt a man behind me, leaning in. I felt him pressing his erection into my back, his labored breath hot in my hair. I felt him and did nothing. I stood still, my body hardening at the edges.

"Hey," I heard a woman say. "That guy behind you. Is he with you?"

"No."

"Hey, buddy," she said to him. "I see what you're doing to that woman."

The guy stepped away from me, eyes glued to the dirty subway floor. Neither he nor I said anything. Both of us guilty. Him for violating my body, me for not being able to conjure up disgust for the intrusion. This woman on the subway having to defend my body for me. Thinking it had boundaries, that the touch of a stranger was somehow new or unacceptable to it.

"EVER FRENCHED A BOY?" LINDSEY ASKED ME AS WE SAT ON her front porch, sipping Shirley Temples. Lindsey had moved to a new neighborhood and new school, so we didn't see each other much anymore.

"Frenched? You mean, like, spoke French to? Yeah, I guess."

"No, I mean, like, *Frenched*. Like, let a boy put his tongue in your mouth."

"Ew, no."

"I have."

"Really?"

"Yep." Lindsey folded up the bottoms of her jeans and stretched out her legs to warm them in the sun.

I tried my best to fold up my own pant legs.

Lindsey shook her head. "Why are you wearing that?"

"I don't know." I was wearing a black velvet pantsuit from a vintage shop. My mom had said it was straight from the 1970s. I loved the sheen of it, the rhinestone buttons, and the idea of being inside another era.

"What's Frenching like?" I asked quietly.

"It's very sexy."

I didn't know what she meant, and I had a feeling she didn't either.

"Does it feel good?"

"Oh, yeah . . . I mean, I don't know. There's a lot of spit, but you just have to get used to that."

Lindsey slipped off her sandals and lay back on the porch, knowing she had my full attention. I pictured her enveloped in a boy's arms, lips touching, mouths open, tongues meeting. I suddenly felt small and unspitworthy. At twelve, Lindsey was starting to grow into a woman. I could barely make it as a girl.

THE FRENCHING WOULD HAPPEN A COUPLE OF YEARS
later at a birthday party at my friend Olivia's house. I didn't know
anyone but her. I nervously squeezed my way through clumps of
teenagers until I found a batch of home-brewed beer in the basement fridge. I drank four in a row, the bubbles tickling my throat
on the way down. I drank until I could ooze into conversations
with strangers without my whole body clenching. I drank until
the floors slanted, until voices became echoes, until my own voice
became thick as stew, until the blue corduroy couch called me to
its lap and I staggered over to the splendidly plush cushions and
passed out.

I woke to a body on mine. Slippery lips, a wormy tongue. Smashing
Pumpkins' *Siamese Dream* playing through the speakers.

I passed out.

I woke to hands on my breasts, hands sliding to my waist.

I passed out.

I woke up alone, nauseated and dizzy. I stood and staggered
to the hall to find Olivia and tell her I'd kissed someone. I was
excited to tell her. I was excited that someone actually wanted to
kiss me. I didn't think of what more could have happened while
I was unconscious. I didn't think it was weird that I wasn't really
awake for it because lots of things had happened to my body when
it was sleeping.

Later, I learned someone passed around a Polaroid of me mostly
unconscious with the boy and his hands all over me. I learned
there were many witnesses. That some of Olivia's friends thought
I was a slut.

I wondered if the boy liked me, if I'd ever see him again. I obsessively worried that with my defenses down, he'd seen the ear and thought I was a freak.

I spent days trying to piece together the night, to bring back the feel of his mouth on mine. To make it romantic.

Fourteen

DURING THE NEXT SURGERY, MORE SKIN WAS TAKEN FROM my lower abdomen and spread across the ear for contouring.

A couple of days later, I woke with a sudden urge to pee. I pulled myself up and sat on the edge of the hospital bed. The slowness of my body was grossly out of sync with the emergency of my bladder.

I fiddled with the tubing of my IV. I hated the IV. I hated being attached to tubes attached to machines attached to the wall. I hated the sudden and random beeping that would splice the air, jolting my shoulders back in surprise. I hated feeling the stiff needle between the skin and bone of my hand that forced my movements to be measured and cautious. And I hated carting the thing with me to the bathroom when I really had to go.

I worked as quickly as my body would allow, rearranging the tubes, slipping off the bed, and crouching on the floor to unplug the IV. Once freed from the wall, I guided the machine down the hall toward the bathroom. Its squirrelly wheels caused it (and me) to zigzag in all directions like a broken grocery cart. I did my best to squish my legs together as I walked. The pee was coming. I was running out of time.

I let go of the IV, praying the tubes would drag the rest of the machine along behind. It worked for a few seconds; then a sudden burn in my hand. I looked down to see I'd inadvertently yanked on the tape securing the IV to my hand and the needle was loose, slanted sharply to one side under my skin. A bubble of blood formed on the skin. I didn't know if it was the unexpected blood, or the unnatural angle of the needle sticking out of my vein, but I was overcome by a wave of dizziness. I thought of my dad, pale and sliding down the wall when the doctors had suctioned out the blood from my ruptured eardrum all those years before.

"It's okay," I whispered, unsure if I was talking to me or the image of my dad fainting.

With urine threatening to escape me any second, I took a breath and pulled the needle out of me. I instantly felt light and floaty, like a balloon escaping the grip of a bratty child. With a surge of energy, I ran to the bathroom. I knew I'd get in trouble for pulling out the IV, and I knew I'd have to get another needle inserted, but all I could feel in that moment was the euphoria of relief.

"DAD WANTED TO BE THERE MORE," MY MOTHER SAYS decades later when I make a joke about him avoiding the hospital because of all the blood.

"He would have put up with the blood, you know that," she says.

"You're right, I know."

"And he drove up to Toronto after each of your surgeries. Stayed the night in a hotel. We couldn't afford to both take time off work, so he could only stay the one night. You were pretty groggy after surgery, so you probably don't have many memories of him there. He did come, though."

I remember. Mostly fragments: him feeding me ice chips to soothe my dry mouth, him telling me silly jokes to try to make me laugh. The din of his melodic voice as I drifted in and out of sleep.

There is one memory that has stayed whole.

"I remember this one time," I say to my mother, "I can't remember what surgery it was. I was feeling really sick. And Dad sat by my bed and read me a Taffy Sinclair book."

"Taffy Sinclair?"

"You know, those preteen books I used to read about a pretty blond girl that a bunch of other girls didn't like. I remember Dad reading the whole book to me while I lay in the bed with my eyes closed. He did different voices for each of the characters."

"I remember that."

"It was quite entertaining."

"It was, yeah." My mother pauses. "He did the best he could. We all did."

"I know."

MY COUSIN LUKE STARTED STAYING WITH US FOR MONTHS at a time while his parents were out of town. One day, not long after my surgery, he and I were watching *The Wonder Years* in the living room. At the commercial, he turned to me and asked why I was wearing my hair up.

"To keep the Polysporin on the ear from turning me into a grease ball," I said.

Luke shook his head. "I shouldn't have to look at that thing."

My throat tightened at the shock of his words. I quietly pulled out the hair elastic to cover the ear.

WHEN HE STAYED WITH US, THE LUKE I KNEW WOULD SOME-times disappear. An anger would take over, and he'd become easily disgusted with me—how I talked, how I walked, how I laughed, how I chewed at the dinner table. When the anger took him, he'd sometimes twist his mouth into a crooked smile, and pin back his right ear to mimic the flattened ear Uncle Louie was making.

We weren't friends anymore. When I asked Luke about it, some-times he would shrug and say nothing. When the anger had him, he'd say, "I don't know, you're ugly," in a voice I barely recognized.

Things were changing. Appearance was becoming more and more important, especially for me as a girl.

More eyes, more expectations. Not just from the doctors, but now from my male peers. It wasn't enough anymore to strive for normalcy; now I had to be pretty, too.

My body was doubly problematic: not normal, and certainly not beautiful.

ON A FIELD TRIP TO GOULD LAKE, A CONSERVATION AREA just outside of Kingston, I sat on the small shore while my classmates sat in wooden canoes on the lake, learning how to paddle. My ruptured eardrum made this too dangerous of an activity for me.

The woman in the cowboy hat—our Gould Lake guide—told us that after the canoeing, she'd be giving us each a special surprise to take home. To ward off the ache of being left out, I spent my time on the sand trying to figure out what the surprise could be. Fresh maple syrup? A bird whistle? Our own little turtle?

After the paddling lesson, we all made our way to the barn, and the woman in the cowboy hat showed us slides of brightly colored plants that grew in the area. Marigolds, snapdragons, geraniums. Orange, purple, red.

"And now," she said, "you get your very own plant!"

I stifled my smile when I realized that none of my other classmates seemed excited.

The woman in the cowboy hat handed each of us a Styrofoam cup filled with soil. We got to choose which flower seed we wanted. I took the longest to decide, eventually settling on a marigold because it had the same name as my favorite character on the *Polka Dot Door* when I was little.

The woman in the cowboy hat showed us how to gently press our seeds into the soil.

"That's all there is to it," she said. "Water it once a week and nature will take care of the rest."

Back in school the next day, my teacher made an assignment out of our seed growing. We printed our names on our Styrofoam cups and placed them on the windowsills of our classroom. Every morning, we'd take turns visiting the cups and drawing the progress of our seeds with colored pencils. The first few seedlings sprouted within a week: tiny green veins popping out of the soil. More sprouted in the days that followed. Only mine and one other classmate's didn't sprout.

The other classmate didn't seem bothered by her seed not growing. "Whatever, it's just a stupid plant," she said.

But I was very bothered. I hated that I was so far behind my classmates.

"Sometimes seeds take time," the teacher told me when I showed her my barren cup, my jaw clenched so I wouldn't cry.

More days passed, and the seedlings of my classmates' cups began to shoot up. Stems formed, and then the buds of leaves appeared. My cup stayed brown.

I came back to my teacher, this time unable to keep the tears from rolling.

"It's okay, Kate," my teacher said. "You can just draw from one of the other cups."

I didn't want to draw someone else's seedling. I wanted to draw my own. I started to worry that I'd done something to the seed when I planted it. Maybe I hadn't been gentle enough. Maybe my touch was poison.

"You just got a bum seed. It happens." My teacher tried her best to reassure me, but nothing helped. I was the kid who was unable to make her plant grow. I was the kid with the bum seed.

"We can get you another seed from the grocery store. You can start with a new one."

"What about this seed? What will happen to it?"

My teacher shrugged. "We'll just toss it."

"No." How could I just throw away the seed like it was nothing? It wasn't the seed's fault it couldn't grow.

"Okay, Kate, I've had just about enough of this." My teacher's eyes turned hard. "I've given you lots of options here. Pick one or you'll fail the assignment."

I failed the assignment.

THAT SUMMER, MY MOM SIGNED ME UP FOR SPORTS CAMP at the local university. In the afternoons, we'd head to the park up the street to play soccer, or flags, or hide-and-seek. The afternoons in the park were most fun because we got to run around without the confinement of walls. One day, on our walk to the park, a group of girls from the camp caught up with me and asked me if I'd been in a fire.

A fire? It seemed like a weird question.

"I don't know. I remember one time my mom was making fudge and the stove caught fire."

The girls kept their eyes fixed on me like I was telling them something very important. Then one of them asked, "Is that how your ear got burned?"

My face blazed hotter than the summer sun above us. I realized with my hair in a low ponytail, they could see the bottom bit of the ear, still swollen and red from the latest surgery. Usually people didn't notice the ear unless I told them about it or showed it to them—things I'd stopped doing since showing the boys at school my scars.

"Yes," I said before I fully realized what I was saying. "It got burned in the fire."

I liked the idea of a fire as the cause of the ear being different. That I was once just like the girls asking me this question. I liked the idea that the ear problem wasn't something I was born with. That it wasn't me; it was something that had *happened* to me.

OVER THE YEARS, A LOT OF PEOPLE WOULD ASK ME ABOUT my body. Strangers, people I knew, people I'd just met. I learned to keep the ear well hidden, but sometimes wind or sudden movement would lead to brief exposure. And my crooked mouth, the way it cut to one side as I talked, was harder to conceal. Often the asking would happen at times I least expected it. When I was talking passionately about something, playing a game at a birthday party, singing a song with friends.

I talk of this as though it's something in the past. It still happens. Sometimes when I teach, sometimes when I laugh, sometimes meeting new people.

One day, on a streetcar, I witness it happening to someone else. A man boards, does a double take at a woman sitting near the front. The woman has a rose-colored growth extending from her cheek to her chin. She is silently reading. The man approaches her as the streetcar pulls forward. His eyes are on her face; her eyes are on her book.

He says the words I've heard so many times directed at me. Words that make my stomach clench: *Can I ask you a question?*

I don't know if she feels the same hardening in her guts as I do in this moment. She doesn't look up.

"Hey," he says waving his hand in front of her. "Can I ask you a question?"

She looks up from her book, nods quietly. I know, as she likely does, that his question is the opening to her having to string herself up for his curiosity, to reduce herself to a body she has to explain. To not anger a man twice her size, a man who could ruin her with nasty comments if he became upset.

He sits down next to her. I can't hear their exchange, but he is pointing at her face and I can see her press herself against the back of her seat. She is answering him curtly, her book now in her lap, her arms hugging her waist.

I think to shout at him to leave her alone, but I fear my own intervening will make this woman feel infantilized, or more exposed. And another fear: that he'll turn his attention to my face.

I feel an anger I normally don't get to experience. I'm usually the one being asked the questions and am often too steeped in my own shame, too busy trying to end the conversation as quickly as possible, to think about being angry.

"But it sounds like he was nice about it," a friend will say later when I tell her about the encounter. "He wasn't making fun of her or being mean. And he asked if he could ask about it before he asked about it."

How to get my friend to understand that even the asking about asking is violence? That this woman has likely experienced years of prodding and abuse based on the way she looks and that saying no probably doesn't feel safe. That having a variant body makes others think it's public property, a lesson to learn from. She is not afforded the autonomy to decide if and when and where and how and to whom she discloses information about her own body.

Another friend will get it all too well. "Before I transitioned, people would just come right up to me, say 'Can I ask? Are you a man or a woman?'"

I will nod. As a cis woman, I will never be able to comprehend the full extent of violence inflicted on trans folk, or the fear they are forced to inhabit in a world that is trying to erase them. But I will understand the breached boundaries of being asked to explain a body.

Later in the conversation, I will say: "Sometimes, when I explain my crookedness, they ask to see the ear."

And my friend will say, sad smile spreading: "Sometimes they want to see my genitals."

I will try to imagine the horror of this. A human being, asked to strip and expose the most intimate parts of their biology to indulge someone's curiosity. The assault of a question so ruthless, so inhumane. Although I will know the violation of being asked to display or explain an othered body, I will not know this level of abuse. The questioning is violence. And some inquiries are much more violent than others.

A FEW WEEKS AFTER SPORTS CAMP, I WAS AT THE MALL walking the narrow hallway to the bathrooms. A group of teenage boys lined the walls and were shouting out rankings to the girls as they passed.

"Seven!"

"Three!"

"Nine point five!"

I thought about waiting to use the bathroom until I got home, but I was pretty sure I wouldn't make it. I trekked down the hall, my heart thundering, my head bent low, my hair covering the ear. I was afraid, yet strangely intrigued to know my number, my worth to these boys in the bathroom hallway. As I passed, there was only silence. Maybe they could tell there was something wrong with me. Maybe I was too gross to even get a number.

Then I heard it, faint and bored-sounding: "Five."

Five! Mathematically, and thus unquestionably, average! Brilliantly normal!

ANOTHER TIME AT THE MALL, ELENA AND I WERE HEADING to Zellers to look for the latest New Kids on the Block cassette. I noticed a man sitting on a bench, staring at all of the young girls as they walked by. As we approached, my insides started to squirm. He scared me, the way his gaze bored into the girls that passed him, but I also felt like I needed him to think I was worth looking at, too. I smoothed down my hair and held my breath as we passed. When I looked back, his eyes locked onto mine. He smiled, reached into his pants, and squeezed his penis. My breath shuttered out of me. I could feel my body being suctioned into his eyes. I was so embarrassed I thought I was going to cry. His smile broadened; his head nodded toward the tent in his pants. I whipped back around and laced my arm into Elena's.

WHEN THE ANGER TOOK HIM, LUKE HAD RULES FOR ME. I was not to sit beside him at meals; if we were watching TV, I had to make sure I didn't enter his field of vision and ruin the show for him.

When the anger took him, he'd point out how I walked like a hunchback when I was tired. How my teeth were bucked. How my mouth was crooked. How my feet twirled absently under the dinner table. How my nose sometimes whistled when I breathed. How my laugh was like a screeching record. How my voice was too loud, too high. How the ear looked like rotting fungus.

Starting at age twelve, my life in my house when Luke stayed over became a task of constant self-vigilance on how my body was performing. I kept my hair down so no part of the ear would show. I did my best to extinguish my natural movements to avoid his comments. I governed my body closely. The space I took up became small.

I didn't tell my parents much about it. Sometimes they caught him saying something and would shake their heads at him, say, "Why did you have to do that?" But they didn't understand how bad it was. And perhaps they were afraid of his anger, too.

I didn't stand up to him. He was bigger than me, and when this anger gripped him, I was afraid it would one day drive him to kill me.

Also, I agreed with him. I felt my body was deplorable. I felt responsible for its deviance. I felt guilty for the missing ear and the subsequent surgeries. I deserved the hatred. My body deserved punishment.

I DIDN'T KNOW TO BE MAD AT LUKE FOR THE WAY HE WAS treating me. Instead, I grieved him. He was the boy who rallied the kids in the neighborhood to buy me a satin pouch with a little ballerina inside. He was the boy who used to spend hours with me on the swings, the boy who made funny voices with my stuffed animals to cheer me up when I wasn't feeling well, the boy who taught me how to play marbles.

When the anger wasn't there, the old Luke would come through and we'd go for a walk, or we'd sing along to old episodes of *Video Hits*. This was the Luke I loved. The real Luke.

I believed that if I could just discipline my body enough, I could prevent the anger from overtaking him. I could keep the real Luke with me.

I am very much encouraged by the fact that she wears a ponytail most times and thus, the proof of the cooking is in the eating.

A lie I told to protect my surgeon's feelings. To dampen my guilt for not showing off the ear he was building. Before the appointment, I gathered my hair into an elastic, wind cool on the fresh mess of incisions. I gathered my face into a smile. I became the girl he needed me to be.

Unsent letter I wrote to Uncle Louie to make him like me

THERE WERE USUALLY FOUR OR FIVE MAGAZINES UNDER Luke's mattress in the guest room. I'd pull them out when he wasn't home and flip through the pages of naked women, studying what a desirable body was supposed to look like. The women had glossy, hairless skin, perfectly round breasts, flat stomachs, full lips. Sometimes there were men in the pictures, their hands on the smooth skin of a woman, spreading her legs, pulling her breasts, pushing her down to their groins. The men's faces were always plump with lust.

The most graphic of the magazines showed vaginas up close. In one centerfold, a blond woman lay on a velvet couch, her legs spread, her hands stretching her labia like pink taffy. Her mouth was half open as if sighing. Her eyes distant and glass-blown. The photo was both terrifying and fascinating. I wondered if there was a man on a stool, barking directions at her and bending her into different configurations like the pinch-faced man who photographed me at Sick Kids.

I would sit on my bed with the magazines open and try to maneuver my twelve-year-old body into the poses and expressions from the photos. Legs spread, neck arched backwards, half-closed eyes, sighing mouth.

ONE NEW YEAR'S EVE, WHEN I WAS TWENTY-SIX, I MET A handsome, hairy, soft-spoken man at a party. When he told me he was finishing up medical school, I lost my breath. I was now at an age when doctors could be my peers. I stopped smiling, and made sure my hair blanketed the right side of my head. I didn't want to be called on the flaws that doctors pointed out for me. I got away as soon as I could, claiming I needed to use the bathroom. As I turned around and headed down the hallway, I was aware of the edges of myself, how they cut the air. I was aware of the space I was taking up and the space I was not taking up, the movement of my chest as I breathed in and exhaled, the angle of my feet hitting the floor, how my head sat on my shoulders and how my shoulders tilted forward.

He found me later in the night, after three Alexander Keiths had settled inside me, and asked if he said something earlier that offended me. "No," I said. He nodded and said "Good," a smile breaking on his face, and I realized he wasn't looking at me the way doctors did, but the way men did when they are interested in sex. I decided I would make out with him. For the research. To discover what it was like to be desired by a man in a profession that had continuously pathologized my body.

An hour later our tongues met, our limbs entangled, his hands slid down my back, up my waist, across my neck, and I became two bodies: the one I experienced and the one he measured. The one he measured was the curve of its breast, the bowl of its hips. The hair he wanted to grab and pull. The one he measured was the heat of its mouth on his. The slither of its torso down his stomach. The one he measured was legs split to him. The honey of the inner thigh. The one I experienced moved in angles and illusions, muscles tense, aping the glossy sexuality expected of it.

IN LINE AT THE GROCERY STORE WHEN I WAS THIRTEEN, I saw a headline on a women's magazine cover that read: "Are You Ashamed of Your Body? Six Ways to a Killer Body Image."

I picked it up and flipped to the article.

It told me to quit trying to diet and start eating healthy for good.

It told me to set an exercise goal and stick with it.

It told me to invest in an accessory that makes me feel hot.

It told me to get waxed so I wouldn't have to worry about unruly hairs popping out from my skin.

It told me to buy myself sexy underwear and strike poses in my mirror.

It told me to get over my flab obsession because men love a curvy girl with some meat on her bones.

IT WAS A RUSTED SEED, FLAT ON THE INSIDE OF MY UNDER-wear. When I checked again a few hours later, the seed had grown into a tiny continent. It was real and I couldn't believe it. In the glow of the pink bathroom stall, down the hall from my eighth-grade homeroom, a small slice of "normal" had found its way to me.

I grabbed some toilet paper and rubbed my underwear until the continent smeared at the edges. I took the underwear off, brought it to my face, and inhaled the fresh coppery scent of new blood.

When I poked my legs back through my underwear, I realized I was completely unprepared. No pads, no tampons, no rags of any kind. I bunched up some toilet paper and stuck it between my legs, feeling its rough edges collide with raw skin. I flushed and went to the mirror, trying to detect any changes in myself. Pale face, big teeth, half-baked ear. Nothing different except the new scratchiness down below.

"Womanhood," I whispered, the reflection mirroring the word with odd mouth shapes.

Fifteen

ANOTHER SURGERY. FURTHER EDITS TO THE RIGHT EAR IN
an attempt to match the left. Cartilage removed from the back of the
left ear to set it closer to my skull, in an attempt to match the right.
A bandage encompassing both sides for almost a month, hugging
my head like an old-fashioned boxing helmet. My hair siphoned
through a small opening at the top of my skull, some of it sticking
straight up, some of it cascading over the bandage. My hearing further muffled under layers of gauze and tape. I stayed in my house as
much as I could to avoid stares from strangers.

After two weeks at home, I started to feel restless. I worried about
all the class I was missing. Since my elementary school ran from kindergarten to eighth grade, next year I would be in high school. I didn't
want to fall behind. I was good at school, despite having missed so
much of it. Being smart was something I could like about myself,
something no one could take from me. Every year, a select few students from the eighth-grade class were chosen to take ninth-grade
math at the high school. They were known as Rovers, and I very badly
wanted to be one.

Although the idea was frightening, I began to consider the possibility
of returning to school with the bandage. Aside from the time I showed
my scars to the boys in the playground in sixth grade, most of my classmates didn't really know much about the surgeries. I always returned
to school after bandages were removed, looking similar to before I left.

My mom called my teacher to tell her I would be returning the following week, bandage on. She asked the teacher to talk to the students about it to prepare them for the sight of me.

I heard from Elena that my teacher told the students they had to be nice to me. Elena told me some of the boys were mad that I got to wear a bandage covering my head even though school policy forbade them from wearing baseball caps.

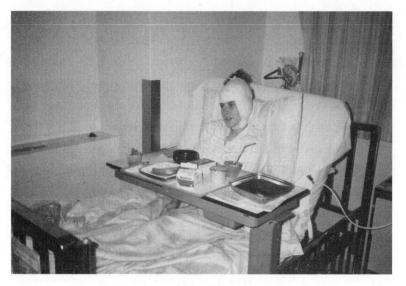

After surgery

I SAT ON THE TOILET IN THE GIRLS' BATHROOM. FIRST DAY back at school and I couldn't get myself to the classroom. The bandage was itchy. I snuck my finger underneath and scraped against the skin. When I pulled it out, it smelled like old pillows and feet. The morning bell rattled through the school and I wondered if I could live within the walls of this little bathroom stall, forever hidden.

Moments later, in front of the bathroom mirror, a reflection. Blue circles hung from the eyes. Her skin was pale, almost see-through. Dark blond hair sprouted like dirty fountain water from the top of the white mess of bandage wrapped around her head. I said to her, "You're ugly." She mouthed it back. I turned on the tap and stuck my hands in the sink. The water ran cold, then warm, then hot. The burn soothed me. My fingers became puffy and red.

Outside my homeroom, my stomach squeezed and released like a greasy sponge. I moved toward the classroom door and heard muffled sounds through layers of bandage: the drone of the teacher making announcements, chair squeaks, students murmuring. I pinched and twisted the skin on my forearm until pink flowers bloomed.

Gotta go in now. It felt impossible, and yet my hands pushed on the classroom door. I entered into a world of desks and eyes. Mostly eyes, peeled and greedy. My footsteps smashed hollow echoes through the hush. I slumped into my seat and focused on the beige speckled floor. When I looked up, no one met my gaze or looked in my direction except to steal glances at the thing on my head when they thought I didn't notice. Then I caught eyes with Jason. The smartest boy in the class. The one with the wheat hair. I tried a small smile. He shook his head, and his lips moved in a series of small directions. A word mouthed to me. "Freak."

THAT NIGHT, ON MY BED IN THE DARK, THOUGHTS RUSTLED like spiders unfurling.

Freak. Freak. Freak.

I didn't want to be in my disgusting body anymore. I rolled onto my stomach and pressed my face into the stuffing of my pillow. I pressed until blotches and squiggles of electric blue danced on the insides of my eyelids. Feet and hands numbed. The world outside flickered and dimmed.

I woke to my own gasping. When I flipped over, the room was dark and glazed with a ghostly sheen that made shadows and outlines seem angelic. In that moment, I was okay. The relief of air was enough.

From a distance, a voice. My mother called from the bottom of the stairs. "Kate, dinner!"

I sat up to a wall of nausea.

"Kate?"

I steadied myself, slid off the bed, and went downstairs.

IT DIDN'T OCCUR TO ME UNTIL MANY YEARS LATER THAT most of my classmates had likely never seen a kid bandaged up like me. I'd had a whole life of seeing child bodies cut and wrapped, appendages missing, parts displaced. It didn't occur to me that perhaps the disgust my classmates directed at me was tangled fear at the realization that things could happen to the bodies of kids, that things could happen to their own bodies. That none of us could ever be truly safe.

MY BLADDER SCREAMED AS I RACED THROUGH THE crowds in the darkened street, searching for a bathroom. I went up side streets and down alleyways until, finally, I saw a door labeled "Women." It opened with a groan. Inside were rows of toilets with no stalls. My feet sank into an inch of yellow-brown water that sheeted the entire floor. I swished through the soiled swamp to find a toilet, but all of them were smeared black. I was losing time. Then I felt it: warmth on my inner thighs.

I woke to myself wetting the bed. The warmth cooled quickly and my underwear stuck to my skin like a slug. I jumped up and ran to the bathroom, the sting of urine meeting wind on my legs.

"Kate, is that you?" I heard my mom ask from her bedroom.

"Yep."

"Did it happen again?"

"Yep."

"Do you need any help?"

"Nope."

In the bathroom, I stripped off my nightie, peeled away the underwear, and sat on the toilet. The plastic seat squeaked as I adjusted my body. I closed my eyes and took in the smell of urine on skin. It was the smell of Brussels sprouts, rusty chains, rotting fruit. It was the smell of garbage.

Years later, a writing teacher will say she doesn't think it's a good idea for me to write about the months I wet the bed when I was thirteen. "People will think you had some deep emotional issues," she said.

ONE LUNCH HOUR, WHILE STILL BANDAGED, I SLIPPED around to the side of the school near the kindergarten area to avoid my classmates, who were doing their best to avoid me, too. A worm was drying on the concrete border that separated the school from the yard. I picked it up. It tried to burrow itself into the lines and folds of my hand, its body gluey and awkward. I felt a tug on my coat and looked down to see a boy, no older than four.

"You gonna eat that?" he asked.

"Ew, no!" I said, and dropped the worm into the grass.

"Does it hurt?"

"Does what hurt?"

He pointed a short finger at my bandaged head.

"Not really."

"Looks like it hurts. Why are you wearing that?"

I shrugged.

"Were you in a car wreck?"

"No."

"A bad haircut?"

"No."

"Then why?"

"I don't know, to make me normal, I guess."

"Well, you don't look normal."

This struck me as funny. All of these surgeries to make me normal were making me anything but.

"No," I said, giggling. "I'm not normal."

WHEN THE BANDAGE WAS REMOVED AFTER A MONTH OF IT tight and increasingly smelly around my skull, my head felt expansive, like a star exploding light. The world opened up. Sounds previously flattened from gauze and tape came at me plump and full. My hair stuck straight up, still in the shape of the bandage.

Uncle Louie talked about next steps for the ear. Another skin graft, more contouring, but I was barely listening. I was in the sky.

Back in Kingston, my mom and I went to Shoppers Drug Mart to pick up mild soap and Polysporin to clean the now-exposed incisions. She let me buy a Red Raspberry Lip Smacker and Timotei Honey & Herb shampoo. The prospect of washing my hair for the first time in a month was exhilarating.

That night, in the bath, I gently eased myself backwards into the water. I could feel tiny bubbles skate up my neck and into my hair. I massaged in the honey-herb shampoo. My fingers drew small circles into my scalp, and my stiff bandage-shaped hair started to relax. The lattice was breaking, the leftover glue from the medical tape dissolving. I shampooed for longer than I needed, relishing the feeling of my hands in my hair. When I leaned back to rinse, my newly washed hair resubmerged and felt softer than I ever imagined hair could feel. Like velvet, like a river.

THE NEXT DAY AT SCHOOL, MY HAIR WAS CAREFULLY brushed. Although the Polysporin on the incisions slicked it with grease, my hair felt like my own again. I wore my favorite red turtleneck, which fit over my head now that the bandages were gone. With my hair washed, my lips coated with my new Red Raspberry Lip Smacker, and Love's Baby Soft perfume dabbed on my wrists, I felt almost pretty.

Elena and I were making friendship bracelets behind the baseball diamond at recess when Jason and Kyle and a couple of the other boys walked over.

"Hey, Stumpy," Jason said.

This was my new nickname at school. I couldn't quite figure it out, but it had something to do with an idea that I had a stump instead of an ear.

"What is that, lipstick?" he said.

"It's gloss," I said.

"Why are *you* wearing that?"

"Because she wants to?" Elena rolled her eyes at him.

But I understood immediately. The emphasis he placed on *you*. I didn't deserve to wear lip gloss because I wasn't really a girl. I was a freak, a thing with missing pieces.

"Looks like shit, Stumpy," Jason said, shaking his head.

Kyle laughed and slapped Jason on the back.

I kept quiet, doing my best to look unconcerned while holding back the wave of tears that threatened to break through. Like Luke, Jason was angry at me—and I didn't know what to do to make it stop.

It didn't help when later that day we found out we had both been chosen to be Rovers. Jason and Stumpy would be sharing the backseat of a cab from the high school every morning.

MANY YEARS LATER, A CASE IN WHICH A QUÉBÉCOIS BOY with a facial difference is mocked by a comedian will make its way to the Supreme Court of Canada. The boy will be the same age as I was when I walked into my classroom with bandages around my head.

The comedian will joke that he thought the child was dying and thus defended him to his friends . . . until he learned the child was just "fucking ugly."

The comedian will joke that he tried to drown the kid in a pool.

The child will be bullied relentlessly by kids using the comedian's material as ammunition.

What would it have been like, I will wonder in horror, if the boys in my eighth-grade class had had the added support and armory from an adult celebrity who joked about murdering me?

The child will get so depressed, he'll want to die.

The Supreme Court will rule the comedian's jokes were not sufficient to claim a breach of dignity, to count as discrimination as defined in the Quebec Charter of Human Rights and Freedoms.

An adult publicly mocking and ridiculing a child with a facial difference, for the purpose of paid entertainment, will be allowed in this country.

An adult joking about murdering a child with a facial difference, for the purpose of paid entertainment, will be allowed in this country.

THE LESION

The plastic surgeon's office, the one the dermatologist refers me to, leaves a voicemail to book an appointment for the removal of the lesion on the earlobe. "Dr. Conrad has some availability next week . . ." I reach for the translucent orange bottle imprisoning my tiny white antidepressants and shake. "Please call us back at your earliest convenience." I like the sound of the pills rattling against the plastic. It reminds me of the maracas we used to shake in music class when I was in elementary school.

After I listen to the message, I type "Dr. Conrad Plastic Surgeon Toronto" into Google, find a photo of him from his clinic page. Light brown curly hair, large Adam's apple, ears slightly pinked. A wide smile climbing a little too far up his cheeks. White glittering teeth. He looks like a prematurely aged teenager.

I picture him pulling my hair back, leaning in to get a better look at the ear. I can smell his cologne, bold and bitter; I can feel his hands pushing into the earlobe, his breath on my neck.

I don't call the office back.

I DEVELOPED A METHOD FOR CHANGING IN THE LOCKER room after gym class. I would go to the far corner, away from the rest of the girls. They would slip off their gym shorts and rummage through their bags for pants, their pastel underwear peeking below their T-shirts. I'd have my pants and turtleneck ready before taking anything off, prepared for the quick exchange. I'd bend to the right, slip off my shorts, and quickly yank up my pants. I'd pull my arms out of my T-shirt sleeves, grab the turtleneck from below, and snap it up and over my torso in one swift movement. I'd keep my eyes vigilant to make sure nobody saw my buttocks, my stomach, my ribs—the places my skin was broken.

A few years later, I joined the high school rugby team and learned that my ears had to be taped solidly to my head to protect them during the intensely physical game. The ritual involved pulling up our hair in high ponytails as our coach wrapped layers of tape around our heads to secure our ears.

Before our first game, I snuck away to the locker room during warm-up and taped my own head. When I came back to the field, my head sloppily taped to ensure that no signs of ear deformity were exposed, my coach gave me a funny look. "Right," I said, "you're supposed to do the taping." I would say this before each game, when I would return to the field after taping in the bathroom.

Eventually my coach stopped giving me weird looks and accepted I was either extremely forgetful or a little strange. Both of which felt better to me than being a freak.

DURING ART PERIOD, OUR EIGHTH-GRADE TEACHER SOME-times let us listen to our Walkmans—a luxury that none of us took for granted. We'd pull on our headphones and disappear into our own worlds of music, bobbing our heads and sometimes accidentally singing a line out loud. One time, when we were pencil-sketching a box our teacher had placed in the front of the classroom, Jason leaned over my desk to get a better look at what I was listening to.

"Metallica?" he mouthed. His eyebrows scrunched together.

Metallica. The newly released *Black Album*. I wasn't normally one for heavy metal, and I knew I didn't look the part. My flowered blouses and high-waisted jeans dangling two inches above knobby white-socked ankles were much more in line with my waning New Kids on the Block obsession than my new love for metal.

Jason was shaking his head. I knew what he was thinking. Luke had already told me that I didn't deserve to listen to Metallica. That it made me a poser, a wannabe who was dragging down the overall coolness of the band, ruining it for those who genuinely deserved to listen.

The popular song from the album at the time was "Enter Sand-man," but my favorite song, the one I'd listen to over and over, was "The Unforgiven." I couldn't explain why I liked this song so much. It inflated my bones. The starting riff pulled at me, each note heavy, weighted, burrowing its way into my chest. Then James Hetfield's voice hammering in, a controlled rage of verses supported by gritty guitar and cracking drums. It was a voice of frustration, rough and throaty, scraping me from the inside. A voice that relinquished to a soft chorus, a minored melody of quiet acceptance. The lyrics spoke of a kid who didn't belong and succumbed to hiding within himself. Or at least that's how I interpreted it. The song was beautiful to me in the way that a dead butterfly was beautiful: with weight to it. A sad beautiful.

I kept my eyes on my half-drawn box, feeling a slight pang of guilt for listening to a band that was out of my league. Jason leaned back into his chair, pointed at my Walkman, and whispered something to Kyle. I turned up the volume until Jason and Kyle became nothing more than jerky movements and twisted mouths, completely out of sync with my song.

ONE LUNCH HOUR, A GIRL IN MY CLASS TAUGHT A SMALL group of us how to faint. It was a combination of taking long deep breaths and then having another girl press hard against the middle of our chests.

The spark of pain radiating to my missing rib cartilage felt good, warranted. A punishment for this body that refused to behave.

Every day for a week, we found our way to the far wall of the school, hidden from the rest of the schoolyard. Near where I'd met the kindergarten boy months before. We took turns pushing into each other's chests, waiting for the numbness to take over.

One of the girls had recently acquired the name Thunder Chunk and had found a tub of powdered SlimFast on her desk. Another girl had told me that her uncle had once licked her privates until she'd peed. Another girl had recently developed large breasts that required a bra that the boys snapped when she wasn't looking.

All of us girls against the red-brown brick of the school's outer walls. Taking turns fainting. Trying to disappear, if only for a moment.

"DO YOU THINK THAT MAYBE YOU DID SOMETHING IN A PRE-vious life to make your ear that way?"

Lindsey was becoming obsessed with the idea of past lives. Who was she before she was her? She'd have dreams and inklings and decide that she must have been a war veteran or royalty in a different life.

I wasn't so big on the idea of previous lives, especially after she asked me this question. Was I being punished with a missing ear for things I'd done in another life? Was I once a bad person? Maybe I'd been that guy my Sunday school teacher once talked about who got his ear cut off when trying to arrest Jesus? Or maybe I'd been van Gogh and cut off my own ear in a fit of psychosis?

I didn't like the idea of previous lives because I didn't want further confirmation that the missing ear was my fault. I already believed it was my fault—that a person deserved the body they were born into. Those with pretty, perfect bodies were simply better people, worthy of good things. Those with bodies like mine were just not quite good enough to deserve a perfect vessel and were thus required to spend their lives atoning for it.

This was not an idea I came up with. This was an idea that lived in the air all around me. An idea I breathed in every day. That my body was a reflection of who I was as a person.

Lindsey needed to believe I'd done something in a previous life to deserve the body I got. She needed a reason. Because then, as long as she was good, she'd never have a body like mine.

MY MOM AND I WERE IN A SECONDHAND STORE THUMBING through bridesmaid dresses, looking for an outfit for my eighth-grade graduation coming up that spring. I headed to the fitting room with a load of candy-colored dresses drooping over my arm like a deflated rainbow.

"Let me know how they look," I heard my mom say as I closed the fitting room door.

I wiggled into the first one—a grape-colored satin gown—and stared into the mirror. The chest area was too big. I took off my socks, rolled them into balls, and stuffed them into the dress where the breasts should go. I swept my hair back away from my face and clasped it at my neck. I tilted my head to the left to look at the latest incarnation of the ear on my right side. The glob of fleshy skin that was supposed to be the earlobe was a slightly paler pink than the rest of it. A thin white scar snaked down my neck.

A line from Stephen King's *Carrie*, a book I'd finished just days ago, popped into my head. *This is the girl they keep calling a monster.*

"Kate? How is it?" my mom asked through the door.

"Not good," I said. "The purple one doesn't fit properly."

"I'll keep looking. We'll find you something that'll make all the boys' heads turn!"

I peeked out of the crack separating the fitting room door from its frame. My mom was raking through more dresses. Looking at her determined face, I felt bad for her. She had no idea I was a loser. No idea that the boys called me Stumpy and were disgusted by the sight of me. *This is the girl they keep calling a monster.* How could I tell her that there was no way a dress, no matter how beautiful, could make me feel like less of a freak?

"YOU WOULDN'T TALK ABOUT IT," MY MOTHER SAYS, YEARS later. "I tried to ask you about how you were feeling with the ear and how things were going at school, and you'd just shrug it off and say everything was fine."

"What would you have done if I'd told you everything was awful? That I hated myself so much that the thought of dying, the thought of exiting my body, was comforting? I knew you didn't want to hear that. I knew you couldn't." I know I'm being unfair.

"I don't know, Kate," my mother says quietly. "I don't know what we would have done."

"I'm sorry. That's not what I meant to say. It all felt overwhelming; things at school, things at home. I couldn't look at it directly . . . It was too big. And there wasn't anything I could do about it. My body, the surgeries, my classmates. And then Luke."

"Dad and I didn't know how bad it was with Luke."

"It was bad."

"I know. God, I'm so sorry. You always seemed so strong. He was having these learning problems in school and just became angry all the time. And he was so scared for you. I remember, when he was younger, he worried about you when you had surgeries. I think on some level he felt guilty that it wasn't him. Lots of complicated feelings. We felt if we punished him, he'd end up hating you more."

"But instead, I hated me more."

IN MY MID-TWENTIES, BACK AT MY PARENTS' HOUSE, BROKE and writing my grad school thesis, I saw Luke. He was also back at home after finishing grad school, and my parents had him over for dinner. After, we found ourselves alone in the living room, our hands around hot mugs of jasmine tea. We talked about the old swing set in the backyard, clouds made of Jell-O, neighborhood forts.

"Then I became an asshole," he said.

"Yeah."

"You know, I think about it. How awful I was. I feel like shit about it."

"Okay."

"Things weren't easy for me either, Kate. I was bullied, too."

"I know."

"They called me stupid. And big ears. And you had these ear problems, too, and I just unloaded it all on you. You were an easy target. You always took it."

"Because I believed you."

"Shit. You shouldn't have. I felt like the freak, and it was easier to make you feel like one than deal with that. You weren't the freak. You were never the freak."

I nodded, said, "Thank you for saying that." But I didn't believe him.

I MET ERIK IN THE PAGES OF A YEARBOOK THAT BELONGED to my friend Olivia. In his headshot, his Catholic school uniform collar was slightly askew and his smile dented at the edges, like a perfect inverted trapezoid. His hair swooped over his forehead and mingled with his furry eyebrows. As he looked at me from the glossed page, his eyes fixed and unflinching, I decided to fall in love with him.

I knew I would never meet him, so I wrote stories about us instead. We were both forsaken by our families and started hanging out with a group of kids under a bridge. When we saw each other for the first time, an instant love sparked in the space between us. We met night after night in the pages of my story. We held hands, we smoked cigarettes, we Frenched by a stream. He read me poetry, his eyebrows arching and flattening as he spoke. We decided to run away to Paris together so we could be writers like Gertrude Stein, like Ernest Hemingway, like Jim Morrison.

But before we could go, and before he could know about my damaged bits, I got gravely ill with some sort of disease that made me look beautifully frail and incandescent. He stole me from the hospital where I was trapped and dying, plucked out the tubes snaking into my body. He took me back to the bridge where we met; he laid me by the stream. We professed our love to each other, and I quietly and beautifully died in his arms. The stream water gurgled sadly. He kissed my dead lips and his tears made my dead face glisten.

The end.

ON A FAMILY TRIP TO NEWFOUNDLAND, I WAS WALKING WITH my parents down a hotel hallway when I noticed a pack of boys ahead—gangly bodies, large Adam's apples, cracked voices. Teenagers. My dad passed them first, then my mom. Then me. I kept my head down to avoid catching the usual looks of disgust I got from male peers. I passed unscathed, but then I heard something high-pitched and shrill. A whistle. When I turned back, I saw something I had only seen in Luke's magazines: eager eyes, wide and shiny. On me. I flushed and looked away quickly, explosions erupting in my stomach.

In the hotel room, I headed to the bathroom and stared at myself in the half-length mirror. Wavy, sun-kissed hair cascaded past my shoulders. My Vuarnet T-shirt gave a slight hint of my newly acquired curves. I tucked back the T-shirt and tied it in a knot at my spine to fully reveal my form. Burgeoning breasts peeked through. *Maybe I don't have to be ugly.* This thought both thrilled and frightened me. I flung off my T-shirt and shorts, rifled through my suitcase, and found the sundress that my mother had bought me for the trip. I slipped into it and raked my fingers through my hair. I went back into the suitcase and pulled out a tube of mascara I'd recently got at the mall with Elena. I slicked my lashes until they looked like the spiky legs of dead flies.

At the hotel restaurant, we sat in front of a wall of windows that looked out onto a parking lot. As I sipped my water, I noticed my dad looking outside.

"They're looking at you," he said.

I turned to look and yanked back quickly. It was the boys from the hall, standing not fifteen feet away from the window, bouncing around a tennis ball and smoking. I looked again. One of the boys looked back and smiled. I looked away.

The server approached the table. My parents ordered their meals and then it was my turn. I couldn't focus on reading the menu.

"They have chicken fingers here," Mom said. "Your favorite."

"I'll think I'll just have a salad."

"Are you sure?" my mom asked. I'd never ordered a salad in a restaurant before.

"Yes." I didn't want the boys outside the window to see me stuffing my face with bread-fried chicken.

When the food arrived, I could barely eat. I carefully plunged my fork into the leaves that were my dinner, arched my neck backwards like the women in Luke's magazines, and chewed discreetly, desperate to pass as a pretty girl, desperate to not be the freak for once. I kept flicking my eyes to the window to see if the boys were still there.

It was not until much later in my life that I realized that freak and pretty are not that different. Both require an audience.

In Newfoundland

NEAR THE END OF HIGH SCHOOL, MY FRIENDS AND I WOULD go to the bar with fake IDs. At the bar, I was someone else. Someone sexy, desirable. With the lights low, my hair meticulously curled, my face glazed in makeup, it was easy to hide myself. I hated the loud synthetic music, but on the dance floor, I moved my hips, mouth half open, and ran my hands along my torso. Men grabbed me, rubbed themselves against me. I kept dancing. I let them stretch me like bread dough, my arms pulled up, my arms pulled down, my arms around their waists. Some of them sighed into my neck. Some of them told me to come home with them. Before the lights came up at the end of the night, I'd be gone. This was how I stayed the fantasy.

Not long after that, I met my first real boyfriend. For two and a half years I hid my ear from him. Every time we saw each other, I'd think about how to break up with him before he accidently saw something I didn't want him to see. One night, lying on his couch, my shirt crept up, exposing a white scar on my abdomen. When he asked about it, I said, "Appendicitis."

After the first six months of dating, he told me he loved me and I became a blizzard of emotions: There was delight at hearing those words directed at me from a boy. There was also guilt—for tricking him into thinking I was beautiful, for tricking him into loving me, for hiding myself so well with makeup and long hair and sex in the dark. And there was the sadness of feeling that no guy could possibly love me if he really knew what I was.

After two and a half years of dating, I couldn't keep up with the hiding; it made me sick with anxiety every time I saw him. My only option, I believed, was to end it.

"We have to talk," I said one night in his basement.

He could tell by my tone that it wasn't something he was going to like. His bottom lip slipped back and forth between his teeth. He spoke: "I don't know what you're going to say, but can I just say that I love you before you start?"

I wanted to reach out for him, take his hand. I wanted to take this last *I love you*, hold it to my face, feel its warmth. I wanted to tell him I loved him, too, because I did. I wanted to tell him as I had so many times. But I couldn't.

When I opened my mouth to tell him it was over, something else came out instead.

"There are things you don't know about me," I said.

"What do you mean?"

It came out in a gush of words: the missing ear, the surgeries, the bullying. It came out like vomit, raw and torrential. When I was done, I felt relief, but also like I had no skin.

He placed his hand on mine and I pulled it away, preempting the disgust I thought he must now feel for me.

"I think I'm going to go." I stood.

He pulled me back down, pulled me into him. My mouth pressed against his T-shirt.

"It doesn't matter," he said. "Any of it. I love you."

I didn't believe him. There, with my face in his chest, I felt hideous. His mouth found my neck, and I crunched my shoulders and turned away. How could he love me now? He took my face in his hands and I felt his gaze. I looked down.

"Hey," he said.

I didn't respond.

"Hey," he said again. "Why won't you look at me?"

"I just can't."

We dated for two more years. We never spoke of the ear or surgeries again, and no matter how many times after that night he told me he loved me, I never fully trusted the words.

ON MY NEXT VISIT TO SICK KIDS FOR A CHECKUP, I WAS greeted by a man wearing a *Muppets* T-shirt under his white coat.

"I'm Dr. Moppit," he said, tapping his shirt with a smile. "I work with Uncle Louie."

He extended his hand and I shook it limply.

"All right then, let's see how things are doing, shall we?" He pulled my hair back and pushed his fingers into the newest incisions. My shoulders crept up their familiar route to my neck.

"Ah, good, things have healed nicely!"

"Okay." I hopped off the examination table.

"Hang on, there's something else I want to talk to you about."

I hopped back up onto the table.

"How old are you now, fourteen?"

"Thirteen."

"Thirteen. Right. I was looking over your chart and your pre-op pictures and . . ." He squinted his eyes slightly. "What do you think of your face?"

"I don't know." I did know. I hated every part of my body. I certainly didn't want to have a conversation about my face with this man.

"See . . ." He took a pen out of his front pocket. "Your face. It's asymmetrical." He ran the pen along my cheek and chin.

My breath tangled in his words.

"Like if you take a look at your mouth, it's crooked here when you talk." He pressed his pen into my bottom lip.

"I know," I said.

"We can fix this, all of it, in a few surgeries. Harvest some nerves from your thigh and stick them in there." He put his pen back into his pocket. "It'd give you a pretty smile."

Minutes later, in the bathroom mirror, I spread my mouth wide and pulled the crooked bottom lip down to see what I was supposed to look like when I smiled. Although people had pointed out my crooked mouth before, having a doctor—a trained professional—point to it made the mouth feel all the more deplorable.

I felt the burn of holding back tears. I knew it was never going to be over, this renovation of my body. There was always more to fix.

SEVERAL YEARS LATER, A TECHNOLOGY WILL EMERGE THAT brings me back to this moment in the mirror when I pulled my lip down in the hospital bathroom to form a perfect smile. Back to the moment my first surgeon pushed a plastic ear into the side of my head. In a time when much of young lives will happen through a screen, when images and avatars will stand in for flesh and bone, social media platforms will launch filters that alter the face. The mirror of my childhood will become a screen displaying hyperrealistic images of society's idealized version of what girls/women should look like: flawless skin, enlarged eyes, pronounced sex-flushed cheekbones, plump lips. A generation of young girls given an instant surgical consult written right into their faces. I will be sickened by the altered image presented of me: a caricature of a sexed-up woman, unrecognizable to myself. A culture claiming to promote body positivity continuing to lure women and girls into measuring themselves against an impossible male ideal. The same old shit.

AFTER DR. MOPPIT I BECAME HYPERAWARE OF MY CROOK-edness. My smile was no longer an instrument to express happiness, but a thing to further detract from my physical appeal. I made my smile small to minimize the crookedness. I smiled with my mouth closed. I learned to push laughter down and say things like "That's so funny" instead.

And joy, always colored with a touch of shame, got tempered through a smaller and smaller hole in my face.

THE KATE I'M SUPPOSED TO BE WAS PRETTY IN THAT PER-fectly symmetrical way that biologists talk about when they talk about beauty. She laughed a lot—an effortless toothy laugh pushed forth from a perfectly proportioned smile. Her skin was silky and flawless. She made the volleyball team when I didn't; she danced with Jason at the Halloween dance when I sat in the corner of the gymnasium alone. She went to lots of parties when I stayed home and watched *Designing Women*. She and Luke were still the greatest of friends. She wore her hair in a high ponytail. She knew what the breath of wind felt like on her neck.

IN GRAD SCHOOL, I TOLD A GUY I WAS DATING THAT A FRIEND of mine was thinking of getting a breast reduction. His face scrunched sour.

"What?" I asked.

"The scars," he said.

"That doesn't matter, though, right?" I looked at him tentatively.

"Yeah, it does. To guys, at least."

A familiar dread awakened in me. He didn't know about the ear yet.

"What? What'd I say?" he asked when I let a silence stretch between us.

"Nothing."

It took incessant prodding, and an eventual assessment from him that I was "being intentionally difficult," before I lifted up my shirt and showed him the scars across my ribs and belly. Like I did many years ago to a group of eleven-year-old boys.

He was quiet, then said, "Shit. I feel like an asshole now."

I liked that he felt like an asshole. That the shame in that moment was his and not mine.

IN THE SPRING OF EIGHTH GRADE, MY DAD TOOK ME TO THE pet store and said I could pick out three goldfish to come live with me. As I approached the tank shimmering with gold and silver fish, I noticed two tiny, almost invisible creatures darting from side to side. They were the smallest, most adorable things I'd ever seen.

"Those are feeder guppies," the store clerk told me when I asked.

"Can I have them?" I asked.

"Oh, actually, they're food for other fish."

I didn't understand. "You mean to be eaten by the bigger fish?"

"Yep."

"Then we've got to get them out of there."

I took the feeder guppies home and poured them into a little bowl with pink gravel and a clay skull. When their tails started to get ragged, I called the pet store.

"Yeah, sometimes feeder guppies eat each other," the guy said.

"Why didn't anyone tell me?" I asked, prickling with anger. How was I supposed to know that these little creatures, prey to other fish, were also prey to themselves?

"Well, people don't usually buy them for pets."

I went back to the store and bought another little fishbowl so that each guppy had its own space to live in. I placed the bowls right next to each other so they wouldn't get too lonely.

The guppies would meet at the glass edges where their bowls touched and I'd watch their tiny bodies dart at each other, knowing that one body would destroy the other if given the chance.

FOLLOWING A WHOLE SEMESTER OF ROVERING, I WENT INTO my final math exam with an A+. My parents were so proud that they bought me a new calculator for the exam. It was a work of art, each button a plastic jewel—ruby, emerald, amethyst, diamond.

After the final, back in my eighth-grade homeroom, I took the calculator out of my bag and laid it on my desk while searching for a pencil. Jason caught sight of it and grabbed it away. I didn't see the calculator again until the lunch bell sounded and he pulled it out of his desk to show his friends.

"I saw her using this stupid thing in the exam today," Jason said. "They aren't even real jewels." He threw the calculator over to Kyle, who pitched it back to him.

I didn't understand. Not real jewels? Did he expect me to have a million-dollar calculator? I stood at the classroom door, not sure what to do. "I need that back."

"Seriously, what is this? Fake cheap calculator."

"Give it back."

"You like fake things, eh?" Kyle says. "Like your fake ear."

"Give it back."

"Why? It's ugly," Jason said, shaking his head at me, "just like you."

I didn't know how to argue with that.

"You want it?" Jason asked. He cocked his arm and fast-pitched the calculator in my direction.

Then something happened that surprised us all: I caught the damned thing. One-handed.

None of us knew what to do. I held the calculator firmly. The lines in Jason's face moved from soft confusion to anger.

"I can't believe they let you take high school math," he said. "You can't even grow your own ears."

I tightened my grip on the calculator. I felt the plastic buttons press into the base, but I didn't soften my hand. This calculator was mine and he couldn't have it.

Part Three

POSTOPERATIVE

I understand why you look at me the way you do: a boarded-up window with a window painted on it.

Thalia Field

IN NINTH GRADE, I WROTE A POEM FOR MY HIGH SCHOOL'S writing contest. I tried my best to be wordy, dramatic, full of feelings, like the poems I liked to read. The last lines were:

As I lie in pastures of thought, a prisoner of my fate
I desperately try to look over the edge and see what's beyond the gate.

I won third prize and went to a little ceremony in the library where some of the English teachers made nice comments about the poem and handed me a check for ten dollars. The teachers thought the poem was about cusping on adulthood. It wasn't. It was about loneliness.

JOHANNA, A GIRL WITH A GLASS EYE, WAS IN MY HIGH SCHOOL philosophy class. Some people stared at her, and some made jokes about her "crying her eye out" when she teared up in the girls' bathroom. One day, she casually brought up her missing eye during a class discussion. We were silent. What was she doing? Didn't she know she wasn't supposed to talk about it? That her missing eye wasn't even hers to talk about, but belonged to all the perfect-bodied kids that stared at it and talked about it behind her back?

It took me a minute to understand the feeling in my chest as I sat at my desk, a curtain of hair cloaking my ear. It wasn't disgust, or anger, or embarrassment for her. It was envy. At the ease with which she owned this missing eye.

THE LESION

The first time I have a night terror, I'm sleeping beside a man I'm on the brink of calling my boyfriend. I've decided to stop taking the antidepressants and do so without thinking about the possible side effects of abruptly ending medication. The screams come out of me before I'm awake. By the time my almost-boyfriend wakes me, I'm already sitting up, shaking, my voice rough from the screams that continue to escape me. I can't stop. The fear rattling inside me feels worse than anything I've felt before. Like being buried alive, suffocating in a tin of my own filth.

"I need to walk," I say to the sheepish face of my almost-boyfriend. I leave the bed and head to the bathroom. I pass the living room where two of my almost-boyfriend's roommates are sitting with beers. They laugh as I pass. A muffled laugh, like they are trying to keep themselves stoic. One of them says something, but my bad hearing prevents me from deciphering his words.

In the bathroom, I climb into the empty tub, hold my hands to my knees, turn on the water, and, for several minutes, listen to it gurgle down the bathroom drain.

The next day, my almost-boyfriend's roommates look at me, bemused. "How was your night?"

Later, my almost-boyfriend tells me that his roommates assumed my screams were sexual. That he had pleasured me into multiple intense orgasms. Deeply embarrassed, I ask my almost-boyfriend to correct them.

"Why?" he says. "Who cares?"

He doesn't correct them. The story of my terror becomes a story of me being loud at sex, and him being a master of female pleasure.

ON A TRIP TO MAINE WITH MY PARENTS, I SPOTTED A BOOK on a large wooden table at a café-bookstore. The image on the cover was a black-and-white portrait of a young girl veiling her face with what looked like a garbage bag. The title drew me in: *Autobiography of a Face*. I picked it up, flipped to the first chapter, and read. On the page was a teenage girl like me, working at a kid's birthday party, hiding herself behind a curtain of hair. Like I did. Instead of a half-made ear, she was hiding a half-missing jaw. I was startled by how much her feelings mirrored my own. Reading her words, I almost cried. I, of course, bought the book. I read it cover to cover in two days.

I read about how Lucy Grealy underwent multiple reconstructive surgeries—bone grafts, skin grafts, tissue expanders. It was the first time I had read my experience on the page. I nodded as she described the constant promise of a better self, the multiple failed surgeries, her feelings of worthlessness.

"I'd think to myself, *Now, now I can start my life, just as soon as I wake up from this operation*," she wrote. "I knew there would always be another operation, another chance for my life to finally begin."

"Yes," I whispered to her words.

Granted, there were differences between us. She'd had many more surgeries than me, and had also survived childhood cancer. Unlike me, she was a good patient, never cried, was always brave, and loved her doctors. Also, Grealy was a twin, which meant that a flesh-and-bone Lucy She's Supposed to Be existed for her in real space. What was that like, I often wondered, to live side by side with the potential of what she could have looked like?

Despite these differences, I felt that Grealy and I were cosmically connected. I drank up her words, looking for a way through my life. A path. I kept flipping the book over to the author photo as I read. Was she happy? Could I be happy? She looked happy. She looked peaceful, content. Confident.

Maybe someday, I mused, I could be those things, too.

JUST SHY OF A DECADE LATER—AFTER LUCY UNDERWENT many more reconstructive surgeries, and after her dead body was discovered in her friend's bathtub when she was thirty-nine—I find an interview online. Speaking about *Autobiography of a Face*, she says:

The story is actually applicable to any, any thing that comes along and tries to tell you, "This is how reality works. This is how you work. This is how you work in relation to reality." And so many things don't offer us the space in which to say, "No, wait a minute." We don't even necessarily have to know exactly what the answer to the no is, but just the act of saying "No, I, I have a reservation about this, it might actually be something else" is an enormous step which very few people feel allowed to make.

THE DESIRE TO STOP THE SURGERIES WAS NOT SUDDEN. It was something that silently sprouted in me for many years. By the time I was fourteen, I'd had enough. I couldn't be in high school with bandages around my head, or miss class because I was in the hospital. I couldn't be cut into, pulled apart, and put back together anymore.

After fourteen surgeries on the ear—five in Kingston, nine in Toronto—after fourteen permutations of skin and cartilage from age three to thirteen, after fourteen times my body was reintroduced to itself, new agreements made between cells, I was tired. I wanted comfort in my body. I wanted to fully inhabit it.

I told my mother I wanted to stop the surgeries on our way up to Toronto for a visit with Uncle Louie to discuss next steps. This was the only time we ever talked about not having more operations; I didn't even know if I was allowed to say no.

She didn't argue with me; she didn't try to convince me to continue. She simply nodded and said, "Okay." I think she was tired, too.

"I DON'T WANT TO DO THIS ANYMORE."

Uncle Louie looked at me, eyebrows caterpillaring inward.

I could feel my guilt of disappointing him creeping in, but I held strong. "I don't want to have the surgeries anymore," I said a little louder.

"We were going to contour the inner rim."

"I don't want to do that."

"But you wanted to get an earring there, right?" he says. "We could try to pull the ear out a bit with another graft. Then you could maybe get it pierced."

"I don't want to do that either." I couldn't look at him, at the disappointment in his eyes.

He turned to my mom. She nodded.

"Okay then." He pulled my chart onto his lap and flipped through the pages. "We should get you an appointment with Dr. Johnson about the asymmetry of your face. I think Dr. Moppit talked to you about this already?"

"I don't want to do that. I don't want to have any more operations on anything."

Uncle Louie looked to my mother again. She nodded.

"Okay then." He threw up his hands. "No more surgeries." He grinned and I felt the warmth of his smile. He was letting me go. Without the prospect of him cutting into me, I could see him as just a man. A father, maybe even a grandfather. For the first time ever, I wanted to hug him.

"Thank you." I didn't know what I was thanking him for exactly, but I was flooded with lightness and it felt like the right thing to say.

"Sure thing, kid. It's been a pleasure." Uncle Louie patted my back softly.

I kept my smile to myself as it threatened to spread crooked across my face. With lips pressed into a line, the smile shone inward, warming my cheeks, my chest, my belly. I could feel my whole body in that moment.

ON MY FIRST VISIT WITH MY NEW PEDIATRICIAN A COUPLE
of months later, she pulled out a letter from my medical file.

"Just received this from your surgeon in Toronto," she said. "It's
addressed to me, but I don't like keeping things from my patients.
Would you like me to read it to you?"

I said yes, even though I wasn't sure I meant it.

She began reading.

> *I had an opportunity to review Katharine in the Plastic Surgery
> Clinic accompanied by her mother . . .*
>
> *. . . She really should be seen by Dr. Johnson, Craniofacial Surgeon,
> sometime in the next year or so relative to facial asymmetry, not to
> push, but to inform. You might break this to Katharine somewhere
> down the line . . .*
>
> *. . . She is a very pretty young lady and is developing very nicely,
> despite the fact of so many surgical procedures and setbacks . . .*

The words felt like a betrayal. Like Uncle Louie was trying to snatch
my body away again. Why did he write this letter when I had explicitly
said I didn't want surgery on my face? Did he really think my face
was that bad? And what did he mean by the last line—did he really
think I was pretty? Why would he write this in his letter? How could
he say that I was "very pretty" and also that I needed my face fixed?

My pediatrician asked me what I wanted to do.

"Nothing," I said. "I'm not going back."

She smiled. "That sounds like a good decision."

She slipped the letter back into my file.

WHEN I WAS TWENTY-TWO, I APPLIED TO MEDICAL SCHOOL. I didn't know what compelled me to do it. I certainly didn't like blood, or the thought of looking at people under their skin. I didn't particularly like the sciences and I didn't do well in crisis situations. I'd spent enough time in hospitals to know I didn't like spending time in hospitals. Yet, for weeks, I spent hours filling out applications, trying my best to answer the decisive medical school question: *Why do you want to be a doctor?*

I mentioned my own medical experiences, how they gave me an insider's perspective of being a patient. I mentioned I wanted to help people, and all the other stuff people said in med school applications. My words were candy-coated, my sentences breaded and fried.

Why did I really want to go to medical school?

Maybe my reasons were more selfish. Maybe I wanted to be a doctor to understand what had happened to me, to learn the names of all my broken parts so that I could finally find a way to claim them.

"Perhaps learning the body, the science of it, the mechanics," writes Arianne Zwartjes in her book *Detailing Trauma*, "is akin to the psychological quest to hold the dark places open, look into them, deprive them of their power."

In any event, it didn't matter. I didn't get in.

THERE WAS NO CEREMONY OR EXCHANGE OF RITES WHEN the surgeries ended. They just stopped. Ten years, fourteen operations (fifteen, if I count the tube in the other ear), and then nothing. At first it was a relief, like the bloated silence that follows loud music. But in the quiet space where I took my body back, with no one examining or rearranging it, I didn't know how to own it. I was uncomfortable alone with the unfinished parts, the crooked parts, the parts left jagged.

The Kate I'm Supposed to Be was officially dead. I grieved her. I grieved the promise of a full, complete, happy, beautiful, loveable me. I didn't yet understand that the loss of her was not the tragedy of my body. So I grieved. Then I hid. For years, I draped my hair over the ear; I avoided it in mirrors, avoided talking about it, touching it. I turned away. A complete renunciation.

Until I enter a dermatologist's office in my thirties, and the ear becomes too loud to ignore.

Part Four

THE LESION

While our bodies move ever forward on the timeline, our minds continuously trace backward, seeking the shape and meaning as deftly as any arrow seeking its mark.

Lucy Grealy

A PRECANCEROUS LESION ON THE MAN-MADE EARLOBE, A
bottle of antidepressants, and a referral to a plastic surgeon. In my
thirties, I take a trip to the Summer Writing Festival in Iowa City to
escape the thick haze of disquiet that has settled into my apartment,
and to try to piece together the disparate memories of my childhood
that have begun to whirl in my head like lost ghosts.

In a workshop, we write and share our work multiple times a day.
I write about my hospital experiences, about the surgeries, the doctors, the scars, the pain, the fear. There is one thing I keep hidden: I
don't write about what the surgeries were for. People in the workshop
assume cancer, or some other life-threatening disease. When, finally,
on our last day together, a woman asks me directly, I shrink in my
seat. Shame, my insidious old friend, bubbles into my throat. Why
can't I just say it: *I was born without an ear.* I'm mad at my shame,
ashamed of my shame.

"Oh, sweetie, you don't have to tell us if you don't want to," the
woman says when she sees my discomfort.

But I want to. I need to.

"I was born with something missing." I pause, then add, "An ear.
I was born missing an ear."

Saying that string of words out loud makes me lightheaded, like
I could float right out of my body and escape through the ceiling.

I look around the room, desperate for my workshop peers to be
okay with me. The alarm and disgust I expect on their faces doesn't
appear. I see soft faces, kind eyes.

ON THE LAST DAY OF THE WORKSHOP, MY INSTRUCTOR STAYS back with me.

"This writing you're doing feels important," she says. "Have you ever read *Autobiography of a Face*?"

"Yes, oh my God, I love that book."

"You know, I went to school with Lucy Grealy. Right here. We were in the Iowa Writers' Workshop together."

There is so much I want to ask. What was she like? Did she ever talk about her shame? Did she have lots of friends? Was she happy? But I can't get my words together.

"She was complicated, I think," my instructor says. "I didn't know her that well. Her family donated a box of her stuff to the university library. You should check it out."

After leaving the workshop, I head to the University of Iowa library and find my way up to the special collections area. I'm not a student and thus don't have a library card, but when I tell the librarian what I'm looking for, she nods and heads to the back. She returns with a small box. Although it's not heavy, I carry it carefully with both arms and set it gently on a table in the corner.

The first thing I notice is a stack of handwritten letters addressed to someone named Hamry. I don't read the letters—they feel too private—but I lightly run my finger along the imprint of her black writing. I used to imagine meeting her, us becoming fast friends, sharing our stories, understanding each other intuitively. As I run my finger along her writing, I think: *She was here, right here on the page, pressing into the white. This is the closest I will ever physically get to her.*

There is a stack of unfinished poems. I don't read those either, afraid that their incompleteness will make me sad.

It's the stack of photos of her near the bottom of the box that I spend the most time with. In nearly all the photos of her, she's not facing the camera. In some, her hair is draped in her face, her head is turned down and away, like she's looking intently at something on the ground in the opposite direction of the camera. In other photos, she's facing a wall, her whole body turned 180 degrees from the camera, like a backwards mug shot. I know this: the impetus to hide the parts of oneself that feel unacceptable. An ache of grief swells in me. She died alone in a bathtub, body stuffed with heroin, body flushed of feeling. Body now forever hidden away.

BACK HOME FROM IOWA CITY, THE PLASTIC SURGEON'S office calls again to schedule an appointment to remove the lesion on the lobe.

"Hi, Katharine, this is Dr. Conrad's office. Again. We have a spot next Thursday at two p.m., please call us back to confirm."

I don't call them back. I'm still not ready. I worry that I'll never be ready.

MY ALMOST-BOYFRIEND AND I STOP SEEING EACH OTHER and decide to try a friendship. One night, we go out for dinner and drink wine; with our bodies loose, we return to the comfortable habit of kissing. I bring him back to my place. We kiss more; then I pull away, say, "This isn't a good idea." He pushes me down. "Yes, it is." His hands press me into the bed. I squirm under him, try to push him off, a familiar powerlessness taking over.

"Stop," I say. He won't stop. "Stop it!" I'm drowning in the mattress. My legs kick, my body shakes, my knees yank to my core. In my frenzy, I inadvertently belt him in the crotch. "What the fuck," he says. "Jesus." He is angry, but he stops. He stops; I breathe. He pulls his jeans up, heads to the door. I stay on the bed and feel every part of myself sparking.

SICK KIDS HOSPITAL IS A MERE TEN-MINUTE TRANSIT RIDE from where I live in Toronto. I have passed by the building from time to time; its outer walls always tug at my insides, as though calling me in. I haven't been inside since I was fourteen.

Now I come to it with a purpose: to find the pieces of myself that haunted its hallways.

When I step into the entranceway, I'm immediately disoriented. Instead of the white walls, registration desk, and tiny gift shop I remember, I see a Tim Hortons, a Burger King, and rows of dining tables.

I make my way to the hallway on the left, hoping it will lead me to a memory. I pass by a Shoppers Drug Mart, and then the hallway opens up into a grand space, seven stories tall, restaurants on all sides—sushi, Chinese, Greek, Thai. This definitely wasn't here before. The cafeteria I knew was in the basement, warm and pleasant, no extravagant choices in food, but a display of glittering Jell-O in glass dishes that rivaled any rainbow.

I walk down other hallways, other paths, my pace quickening as I get more and more desperate for some semblance of familiarity. And then I find myself in a deserted outpatient waiting room. I know this place: This is where my mother and I waited for hours to see Uncle Louie and his team after surgery. This is where stitches and bandages were removed, and where, once upon a time, a doctor with a *Muppets* T-shirt ruined my smile.

I walk into an elevator and instinctively press 7, both hoping and fearing it will take me to Ward 7C, my old stomping grounds. I step out into more unfamiliarity. I spin around, looking for a ward, my ward. I'm greeted only with more walls and empty hallways.

I don't know the way anymore.

I take the elevator to the basement—the place I always assumed my operations happened. Although there are no signs to operating rooms, there is a sign that catches my attention. "Medical Records Department." I follow the arrow.

"I was a patient here twenty years ago," I tell the man behind the desk. "Any chance you'd still have my records?"

I am hoping he'll say no.

"We have all patient records for the last hundred years," he says. "I can make you a copy. What's your name?"

I FIND OUT THAT THE RECORDS FOR MY FIRST FIVE SURGER-ies on the right ear—those that took place in Kingston—have been destroyed. My mother gives me a small envelope of photographs from the Kingston hospital that she finds in the back of her closet. They, along with the mottled scars on my buttocks and neck, are the only evidence of those surgeries. The specifics of those procedures are lost, and I feel both sad and oddly grateful for not being able to know all the details of what happened.

The file from Sick Kids, however, is almost two inches thick, separated into four sections: "Intake Forms," "Appointment Notes," "Letters," and "Operative Reports." There is also a CD labeled "Photos."

I slip the CD into my computer and spread the photos my mother sent across my desk. Together, there are over twenty images, each chronicling the evolution and failures of the ear. Raw buttock skin, plastic ear, no ear, tissue expander flat, tissue expander full, tissue expander infected, no tissue expander, rib ear, earlobe, higher earlobe, more skin, more skin, more skin. Some photos include my whole head and neck, while others are just disembodied photos of the ear.

I click to enlarge one of the photos on the screen: me at seven years old, hair clipped back, head turned to the side, spotlighting the implanted tissue expander pressing against my skull. The skin is stretched and shiny, slightly bluish, red at the edges. I pull myself closer to the computer screen and inspect the lines and contours in the photograph. The drooped jaw, the eye like an abandoned swimming pool. It takes me a minute to realize that the child in the photograph looks like she's dead.

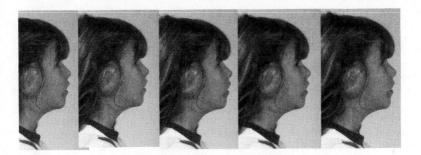

Medical photo: tissue expander infection

"I'M READING THE MEDICAL RECORDS FROM SICK KIDS NOW," I tell my mother over the phone.

"Any surprises?"

"A couple. Not really, though. I can send them to you if you want."

"No thanks." My mother's voice is suddenly sharp. "I lived through it once, I don't need to do that again."

"Okay, that's fair," I say. I'm a little disappointed.

I hesitate before speaking again. "I'm trying to piece together the Kingston surgeries, since there are no records. It was five, right?"

"Yes, five surgeries. And then the surgery to put the tube in your left ear. So six altogether."

"The first one was the skin graft from my buttocks. And then the plastic. Then, the plastic removed. That's three . . ." I wait for my mother to chime in. She doesn't. "Then, another plastic inserted and removed, right?"

I hear my mother sigh into the phone. She doesn't remember. All she knows is there were six surgeries altogether.

"Can you try a little harder?" I ask.

"Kate, I can't remember, okay?" My mother's voice is breaking. She says she doesn't want to remember, that it brings up too much. All of the mistakes. "I feel sick with guilt about it already—the toll these surgeries took on your life. On all of our lives."

"Okay, I'm sorry," I say, my voice softer now. I don't want her to feel like that.

"We thought we were protecting you." When she was growing up, people made so many bad assumptions about those with physical differences. She and my father wanted to protect me from that. "In the end, we didn't protect you, though, did we?"

"I don't know," I say. "I guess I just wish you'd explored other options."

She says they did, or at least tried to. "It's so hard as a parent—you're making decisions that can affect your child's whole life. You want to make the right decisions, but it isn't always clear what those are. Who knows, maybe back then you'd have resented us if we'd done nothing."

She's right. Maybe I would have.

I GOOGLE UNCLE LOUIE AND FIND OUT HE RECENTLY DIED. An online memorial page speaks of his passion, how he dedicated his life to reconstructing broken children. Posts from colleagues, friends, and family speak of his gifted and precise hands, his colorful running shoes, his gleeful laugh, how he used words like *rad* and *awesome* to connect to his patients.

I don't remember his colorful shoes, or his laugh, or his use of the word *awesome*.

I do remember his hands.

There is one post from a former patient. She says Uncle Louie had a huge impact on her life. She says he was her favorite doctor, that he was so kind and caring that she once brought him a cookie cake to show her gratitude for all that he'd done for her.

I feel guilty. I was never grateful. Why aren't my memories of a doctor who I felt cared about me?

There's a post from a father of one of Uncle Louie's former patients. He refers to Uncle Louie as a "magician" who "helped to mold the character" of his daughter. He mentions how when his daughter was married the previous year and "became a beautiful bride," Uncle Louie was foremost in his mind. The father mentions that he even forwarded Uncle Louie a photograph of his daughter on her wedding day to give him the opportunity "to see his artistry in action."

I cringe at the thought of Uncle Louie receiving the photograph and feeling pleased about his artistry.

I cringe at the idea of Uncle Louie performing "magic." Like his impact on bodies was sanitized and spiritual, instead of raw and bloody. Like nothing short of magic could save these bodies.

I cringe at the implication that it was only because of Uncle Louie's "artistry" that this man's daughter "became a beautiful bride"—or perhaps married at all. That her face was his achievement. His legacy.

"This," I say out loud to the computer screen, "this is the fucking problem."

I FIND OUT UNCLE LOUIE IS BURIED IN THE SPRAWLING cemetery across the street from my apartment. This incredible coincidence—this proximity of our bodies—compels me to visit him.

"Why would you want to do that?" my mother asks.

"I don't know."

And it's true: I don't, beyond simply feeling compelled. Compelled to kneel in front of his grave. Not as an act of retribution or closure. More for the hope of understanding him better. For the hope that with our bodies so close, I will know something more about mine.

I live with Uncle Louie in my body. His hands reached into the underside of my skin, places I will never touch. There are parts of me he knew that I can never know. Somewhere in the muscle memory of his body were the stories of what happened to me while I lay unconscious in front of him on the operating table.

I walk through rows of graves, searching for him, wondering if an unexpected emotion will crack through me when I find him. Fear? Anger? Sadness?

I don't find him. I go back the next day and the next. I don't find him.

I READ THE INTAKE FORMS, OPERATIVE REPORTS, ALL OF Uncle Louie's notes and letters.

I linger in the operative reports, the events of me in the operating room that I don't witness. I don't know how many men have had their hands on me. I don't know how many men have seen my forming breasts exposed on the table, have pressed into my groin area to get the perfect cut from my lower abdomen. I don't know the feeling of a scalpel slicing my skin, of a saw cleaving the cartilage of my rib cage. I read the operative reports as though the events happened to another body. But they happened to mine, and there are parts of me, of this body, that know and remember.

Kate Cole-Adams, in her book on anesthesia, writes: "Surgical incision has a galvanizing effect even on an anesthetized patient: As the scalpel enters, her heart beats faster, her blood pressure rises, sometimes she jerks."

Anesthesia does not numb the body from the experience.

"The body still broadcasts its storm warnings along the wiring of the central nervous system and into the brain, but here the signal is blocked or scrambled."

My body couldn't get through. It wasn't allowed to speak. It couldn't move. It had no mind to assimilate the intrusion. It didn't understand the aesthetic expectations placed on it. It only understood the violence.

PLACES I AM STILL

In the examination room, when he presses his fingers into my skin, holds me down to quiet the rustling. When he feeds me anesthetic until my eyes roll back, my body limp and open on the operating table.

At school, at home, when he is disgusted. My body an insult to his idea of Woman. When his names and rules seep into my musculature, fold my bones into an apology. When his anger makes smallness my survival.

In the bedroom, when my mouth is not for words, but soft O's of pleasure. When my drunk-slumped body is his invitation.

A FRIEND INVITES ME TO A CONSCIOUSNESS DANCING PARTY. The internet tells me it's a form of dance that involves moving according to one's inner instincts, one's intuition. A self-expression coming from the core.

When we get there, I'm intimidated. People are sprawled on the floor, bending their bodies into various shapes. I pretend to stretch like I know what I'm doing, straddling my legs, reaching fingers to toes. When the drumbeats start, everyone gets up and starts moving. I don't move. I don't know how to move.

"Just let your body move the way it wants to," my friend says to me. "Anything goes."

I have no idea how my body wants to move. For so long, dancing has been for other people. I stand frozen as people flit and jump around me. I squeeze into myself to keep out of everyone's way.

I soon realize I'm making more of a scene by not moving, so I start to sway. A man in a white tank top and blousy pants approaches and, without warning or permission, brushes himself against me. Suddenly my body knows what it wants to do. It wants to be big.

I stomp, I twist, I fling my hands in all directions so that anyone who tries to get near me will get a finger in the eye. I close my eyes, reach to the sky, and, for a moment, feel myself expanding, filling out my edges. For a moment, I'm full and radiant and swirling with energy. For a moment, I forget my form and what it means to other people, and live fully inside this body of mine.

"I AM PERSONALLY AFFRONTED BY THE MESSAGE THAT I AM only acceptable if I look 'right' or 'normal,' where those norms have nothing to do with my own perceptions of who I am. Where 'normal' means the 'right' color, shape, size or number of breasts, a woman's perceptions of her own body and the strengths that come from that perception are discouraged, trivialized, and ignored."

This is Audre Lorde's response in *The Cancer Journals* when she was encouraged by doctors to get a prosthetic breast after her mastectomy. Although her oppression extends well beyond my own—she, a queer woman of color, living with the aftermath of breast cancer—this quote resonates, gives me strength. I've never had a perception of my own body beyond what others have said of it. But I want to.

I READ THE INTAKE FORMS, OPERATIVE REPORTS, ALL OF Uncle Louie's notes and letters.

Nowhere in the medical file does it address the question of why I had to have the surgeries in the first place. To get inside this question, I have to consult other sources.

"Disfigurement," writes sociologist Heather Laine Talley in her book *Saving Face*, "reminds us of the fragility of the human condition (and human bodies). It is seen as an awful experience rather than a variation on human life. Congenital anomalies are the physical manifestation of the idea that we're born imperfect. We like to feel we have control over things that we don't have control over. It's fear that this is humanity—this is part of the human condition . . . and life is inherently risky."

This societal fear, relegated to my body.

My body, a disruption to order.

My body, disciplined into an acceptable, knowable, controllable form.

A FEW DAYS AFTER THE CONSCIOUSNESS DANCING, I GET MY hair cut by a new stylist a friend recommends.

"He's a little weird," she says, "but he gives a hell of a cut."

When he sits me in the chair, I start the spiel I always give to new hair stylists. I tell him he can't wash my hair because of the ruptured eardrum, but that he can spray it down while I cover the ear with a towel. Before I can tell him that he will likely notice the scars and the physical difference of the other ear, he interrupts.

"Your mouth," he says, "it's crooked when you talk. Just like mine." He stretches his lips into his cheeks for me to illustrate. His lower lip cuts diagonally across his chin. "Bell's palsy, two years ago."

I think this is going to be a moment of comradery. It isn't.

"I can't stand it," he says. "I'm so ugly now." He stares at himself in the mirror, pulling on his lower lip. "I found this guy, though, he's like an acupuncturist/muscle memory/face healer guy. He's fixing it."

"Oh." I can't think of anything else to add.

He spreads his arms above my head. "He can fix you, too! Yes! I'm going to go get you his card." He disappears into the back.

This is not new. Many people I've encountered over the years have ideas for me and my face. Like the woman who, after I met with her and her son to help him navigate his academics, pulled me aside to tell me she was an acupuncturist and could fix my crooked face. Or the osteopath I went to for back issues who shoved his fingers into my mouth and pressed hard on my jaw, in an attempt to fix the crookedness I didn't come to see him for.

The stylist returns with a little yellow business card. "Here."

"No thanks."

I let my blunt refusal hang in the air between us. He doesn't seem to notice. He tucks the card into my purse beside the salon chair. "Just in case," he says.

I realize in this moment I don't feel hurt. I feel solid in my chair. And a new feeling: empathy for this man and his misguided attempts to be helpful. For his tormented feelings about his own face.

"I THINK I MIGHT BE WRITING A BOOK ABOUT THE EAR," I SAY to my mother.

She asks me if that's a good idea. Not the response I was hoping for.

"I don't know, but I just need to try to make sense of it. The surgeries, my feelings about my body."

"Am I allowed to read it?"

"Of course. If you want."

She asks if it will upset her.

I tell her I don't know.

She is silent for several seconds. "Was I a good mother?"

"Yes, my God, yes. The greatest."

"I don't feel like it. I feel like I allowed things into your life that clearly made it so much worse."

I remind her that she always stayed with me in the hospital, even when the nurses thought I was too old to have my mother sleep over. I remind her that she always had a present waiting for me when I got out of the operating room. That she rubbed my back and watched movies with me. That we took walks down to the cafeteria and she bought me Jell-O. That she played crazy eights with me over and over until the stretcher came to take me to surgery. I remind her that she walked with me to the operating room as far as they would let her. I remind her that some of the kids there had no one, and that she watched over them, too. "Through all of it, the pain, the fear, you were there," I say. "In a way, it was our time."

"I wish we could have spent it another way."

Mom and me, searching in the sand

I READ THE INTAKE FORMS, OPERATIVE REPORTS, ALL OF
Uncle Louie's notes and letters.

The story on the page is of a disfigured body and the men who
worked to save it from itself. I trace my finger along the corresponding
story written into my skin: the pink lines carved into my chest, the
whitened smiles on my lower belly, the red square on my buttocks,
the slight cave where ribs once fenced my liver.

Skin, such a flimsy barrier, splits easily to the blade. Each expertly
sliced incision is evidence of a body destabilized, a body thrown into
crisis healing.

Long after Uncle Louie's own body disintegrates into the earth,
these stories, his stories, Dr. Winston's stories, will live on in my skin.
My body will always be part of the legacy of these men.

I don't know how to be mad yet. Only sad. Not just for the physical
infliction of medical intervention, but for how it left me no space to
participate in the meaning of my own body. For how its mystery and
beauty were reduced to definitions. For the inception of the Kate I'm
Supposed to Be, shown to me in the mirror of Dr. Winston's office
when I was a young child. How this imaginary Kate was the driving
force for the erasure of my original body, for the violence done to it.

I read the intake forms, operative reports, all of Uncle Louie's notes
and letters. No longer to understand my body, but to interrogate what
was said about it.

> . . . *we will just keep working away and try to bring her closer
> to a symmetrical, somewhat ascetic [sic] situation the patient
> was prepped and draped in supine position with the head slightly*

tilted to the left the patient the head the wound was closed with vertical mattress prolene sutures we'll just keep working away I had an opportunity to review Katharine right anotia and associated hemi-facial microsomia the patient was then resuscitated the head tilted she's a very pretty young lady and developing nicely in supine position she really should be seen by the craniofacial surgeon relative to facial asymmetry the wound resuscitated . . .

Medical intervention left me with a body I have trouble living in. One that carries all of the breaches to its borders, all of the words and hands and decisions laid on it.

Variant bodies are medicalized bodies are objectified bodies are abject bodies are bodies split off from the people who house them.

I am split. We are split.

What would it take to let our bodies be?

How does an individual say no to intervention when variant bodies are feared, devalued, and abandoned? When the powerful institution of medicine has its fingers so firmly rooted in the business of erasing the unruliness of variant bodies? Of rewriting variance as a kind of sickness?

I don't have answers. What I know is that I wish I hadn't had the surgeries.

MY MOTHER ASKS ME IF I THINK PLASTIC SURGERY IS BAD.
I have no clear answer.

I tell her I don't think plastic surgery is necessarily bad, but I do think it's sometimes a trigger response to body difference. It doesn't leave room for other options. It doesn't interrogate the broader societal implications of erasing body difference. It doesn't allow for alternate visions for having a body.

I know I have a certain privilege. My missing ear and crooked smile are not readily noticeable. I am not stared at or ridiculed anymore. I haven't been demonized, ostracized, abandoned, or killed, which happens in other parts of the world.

Living with a facial difference in any part of the world is hard. Adults with facial differences are regularly harassed in public and in the workplace. They are regularly asked invasive and inappropriate questions. Selfies of those with facial differences are often censored for being "disturbing" or "violent" on social media platforms. And it is notoriously difficult to protect those with facial differences from discrimination under the current laws and codes that govern this country and many others.

I understand why people with body differences want plastic surgery. We can't put the responsibility on individual people to say no to plastic surgery when it may mean a harder life of judgment, ridicule, and violence. Of discrimination that is rarely recognized in legal systems.

But my question is, why does body difference automatically become a medical issue to be fixed? Why is the problem always relegated to the individual body instead of the society in which that individual lives? Why is it my body's responsibility to assuage others' fears?

Fixing doesn't solve the problem. Fixing comes at a price.

I READ THE INTAKE FORMS, OPERATIVE REPORTS, ALL OF Uncle Louie's notes and letters.

My finger maps the story in my skin. My body a more reliable narrator than the pages in front of me.

My scars are sites where my body has been broken into. They are evidence of the attempts at discipline, normalization, at a clean and proper body, a body erased of its deviance. A body cut to behave. "No one will ever have to know," Dr. Winston said to me all those years ago.

But skin becomes feral when cut, puckers at the seams, refuses pretty. Says, *Something happened here. Something happened.*

I CALL THE PLASTIC SURGEON'S OFFICE AND MAKE THE appointment to have the earlobe lesion removed. The night before, I pace my apartment, trying to prepare.

I'll just tell the plastic surgeon that I'm only here to have the lesion removed. End of story. That I'm aware of my crooked smile and crude ear, thanks, and that I don't need to discuss them. Yes, I'm sure there are newer ways to make better-looking ears. Not interested. Fuck him, it's none of his business. Just because he has to look at it doesn't mean he gets to have an opinion on it. Who does he think he is? In fact, if he asks me any questions or says anything about it, I'm going to leave. I don't care. I'll tell him he's an asshole and I'll fucking leave.

I LEAN BACK IN THE EXAMINATION CHAIR IN THE PLASTIC surgeon's office, crossing and uncrossing my arms. The room is unlike any other examination room I've been in before: marbled floors, a large bay window, a leather reclining chair instead of a metal table. A plant. Brochures for laser hair removal and anti-wrinkle treatments.

The door swings open. "Hi there, I'm Dr. Conrad." He looks down at his clipboard. "So, you have a lesion on your ear to be removed."

An oily "Yes" slips out of me.

"It says here you've had reconstructive surgery to build you an ear on the right side," he says, head lowered to my chart.

"That's right, but—"

"And the lesion is on the ear?"

"Yes, but—"

"Okay, let me see."

I pull back my hair. *Only here for the lesion, asshole.*

"Hmmm. Ah, yes. Okay," he says. "How many surgeries did you have?"

"I don't really want to talk about it, if that's okay. Sorry." I hate that I say *sorry*.

He gives me a puzzled look.

"I'm just here to get the lesion removed, so . . ." My voice is smaller, less sure, than the voice I practiced the night before.

He pauses, shakes his head, then takes a small pointy instrument and sloughs off the lesion in one swift motion. The ear roars in response.

He hands me a small tube of Polysporin.

"You'll need to clean the area with mild soap and coat it with this twice a day for a week."

"Clean it?"

"Yes. You don't want it to get infected, do you?"

I most certainly don't. Cleaning the ear, however, means having to look at it, touch it—and I haven't done either of these things in almost twenty years.

ABLUTION: THE ACT OF WASHING PARTS OF ONE'S OWN BODY. To de-possess. To heal. To enlighten. To care for. To bless. To honor. To reclaim.

I stand in front of the mirror, pull back my hair, and look at the ear for the first time in my adult life. The skin is pale and doll-like. Ridges and folds of borrowed tissue and cartilage assume the basic shape of an ear.

I take a good look at this thing that took on several permutations and prompted years of well-intentioned violence. I take a good look and think: *Such an uproar for such a small thing.*

I lean into the mirror until the ear is nothing but color and form roped with silver string scars.

There, up close, despite the imposed intervention, without the stories and the words to name it, I see it. The beauty. The humanness. "I see you," I say, unsure if I'm speaking to the ear specifically.

Scars resist unremarkability. They are also evidence of a body healing itself, protecting its borders. They are evidence of a body loving itself. Perhaps this is what I can hold onto.

I pull out a cloth and wet it with soapy water. I bring it up to my head and press gently, the cloth a soft hum on the ear. When I perforate the tube of Polysporin with the backside of the cap, I feel the release, almost a sighing. I squeeze the tube until the ointment worms onto my finger. The Polysporin melts greasy tears down to my palm. I brush the finger gently against the delicate bulb of skin that is supposed to be an earlobe. A site not unlike the original pebble of skin that once existed there. Before the surgeries.

A tingling, a brightening, a pulse funnels from ear to toe. I am here, all of me.

Part Five

AFTER

*It's all made up. I mean having a body in the world is not to have a
body in truth: it's to have a body in history.*

Anne Boyer

WHEN I'M FORTY-ONE, I PULL A URINE-SOAKED PREGNANCY test in front of my face and read the two pink lines. It's the middle of a global pandemic—an unfathomable event that somehow feels more fathomable to me than being pregnant. I phone my new partner and we are shocked, then thrilled. I had told him mere months before that I wasn't sure about having a baby at my age, but I hadn't told him the real reason for my hesitation: that perhaps my body is, and has always been, too damaged to grow a life inside it.

My partner comes over and serves us fake Champagne to toast with. We clink our glasses and smile big at each other, at a loss for words. I feel the bubbles of fake Champagne fizz in my throat. I am elated. I am terrified.

I PHONE MY MOTHER TO TELL HER I'M PREGNANT. I DON'T believe the words as they exit me.

This is not a mother-daughter moment of happy tears, or of deference for this body I live in. My mother is worried, cautious, tells me it's still early days. I assume I know where this comes from—years of her witnessing trauma in my body, multiple failures, multiple infections; my body a constant battleground. Us always unsure of what it would do.

I say *if* and *maybe* when talking about the pregnancy with my partner. My mother and I barely discuss it at all. My partner interrupts my *if*s and *maybe*s and says, "Kate, this is happening. You aren't 'maybe' pregnant, you are definitely pregnant."

I take more pregnancy tests, unconvinced of their reliability. I leave the used tests on my dresser for weeks and check them each morning to ensure the two lines are still there. My therapist says to "embrace this beautiful thing happening in your body." She gives me mental exercises in which I place my hands on my belly and repeat, "I'm pregnant. I'm pregnant."

I SCOUR THE INTERNET FOR WHAT TO DO AND NOT DO during pregnancy. No sugar, no caffeine, no fish, no Advil, no heavy lifting. Keep stress low, take prenatal vitamins. I'm a seasoned obsessive health eater by this point, and feel strength and relief in believing I can, once again, make up for my damaged body through diligence and willpower.

I avoid "mommy-to-be" websites, with their pastel colors and shiny young women, with their "your fetus is now the size of a blueberry!" updates. Not because of their whitewashed cheesiness, but because visiting those sites makes me feel like an outsider, like I don't measure up to being the kind of woman who could be pregnant.

One afternoon, I notice a drop of blood in my underwear. For three days, I check my underwear several times an hour. Sometimes a dot of blood, sometimes nothing. On the fourth day, I go to the emergency room.

My partner can't be with me because of pandemic restrictions. Before I walk through the hospital doors, he says: "It's going to be okay; I know it." I don't believe him.

Six hours in the emergency room. Blood tests, IVs, and mask-muted conversations with medical staff. Without the visual facial cues and lipreading I rely on to hear, I am lost. Eventually I'm sent for an intravaginal ultrasound. When the doctor approaches me to discuss the results, I'm convinced he's about to tell me I was never pregnant, that it was just some weird hormone anomaly that turned the pregnancy tests positive. That it's all a sick joke.

It's not a joke. He tells me I'm four weeks along, and everything is where it should be. The bleeding, he says, is nothing to worry about. I worry.

AT SEVEN WEEKS, THERE ARE STILL FAINT DROPS OF BLOOD making ominous shapes in my underwear. I go for another ultrasound.

The line for COVID testing at the hospital is all the way up the street.

In the sound-imaging waiting room, we all sit six feet apart, masked. A woman across from me rubs her plump, pregnant belly, and I put my hand on my own belly, wondering if I'll get there.

In the ultrasound room, legs spread, toes curled, I keep my eyes closed as the foreign rod gathers images of my insides.

"Look," the technician says as she turns the ultrasound screen toward me, "there's the heartbeat."

It's a small flashing. A twinkling star in a sky of black and gray. My chest swells with my own flashing heart.

"Thank you for showing me," I say, not realizing this is standard procedure.

After that, I allow myself to trust I'm pregnant. My partner and I begin throwing out baby names to each other.

Theodore?

Felix?

Roxanne?

Olive?

We watch the credits at the ends of movies, a new wealth of potential names. A hope of a baby floating against the black.

THREE WEEKS LATER, I'M TRYING TO PUSH A DEAD FETUS out of me. The toilet bowl turns into a bloodbath.

I spent the previous day in the emergency room alone, trying to decipher the muffled words coming at me from behind the mask of the emergency doctor. Finally, words reached me that I understood: "No heartbeat."

I refused the dilation and curettage procedure that the doctor recommended, still terrified of operating rooms, still terrified of laying bare my unconscious body on a metal table. The doctor prescribed pills instead.

Now, I make my way to the bathroom floor, the rounded bones of my knees press into the bath mat. My mother is on the other side of the door, there with me again, despite having said she didn't think she could go through this with me: watch me bloody and in pain, like she had so many times when I was young.

"Is there anything we can do?" she asks through the wood.

"No."

I feel the squeeze of grief for this little fetus that had flashed its heart at me on the ultrasound screen weeks before. I feel a helplessness that I imagine my mother has felt many times with me, including in this moment here.

I think about the shock of it for my partner, how maybe I have altered his sense of safety in the integrity of bodies, in their predictability.

All of the scars burned into my skin relight, starting this time on the lower stomach, three scars that, days before, I worried would split open as I ballooned with life. These scars ignite, catching fire to the one at my ribs, my neck, my head, the ear.

Here we are again, body. Flamed and split.

Contractions mount from the pills the emergency doctor has prescribed. The pain feels right, a distraction from the thought that keeps bubbling up in me: *You are too damaged you are too damaged.*

Because shame is hard to scrub from the bones.

Because trauma doesn't end; it echoes.

Because I wanted to end this story with me in my thirties, looking in the mirror and loving my body, but that's not how this goes.

Because true stories ping between crisis and resolution, back and forth and back again, peace and horror, peace and horror.

I push. I hold my hand to my belly and whisper, "You're okay." I'm not talking to the dead fetus moving its way through me. I'm talking to this body, my one permanent home. I'm expressing love, like I learned to do in that bathroom mirror almost ten years ago. I express love even though I have a hard time feeling it. Even though loving this body feels like its own kind of constant pushing. But perhaps also its own kind of birth.

Part Six

NOW

Was I ever seen again?

Marie NDiaye

WHEN MY MOTHER ASKS IF SHE CAN READ THIS MANUSCRIPT, I don't print it off and mail it. I don't send it over email. Instead, we set aside an hour each night on the phone and I read it to her. It feels important for me to share it with her in this way. Perhaps to gauge her reaction in real time. Perhaps to hear my own voice tell this story of my body.

Every day for a week, I phone and read to her. Sometimes we cry, sometimes we laugh, but mostly, it's just my words, traveling through the phone to meet her silence.

She has opinions, clarifications, things she didn't know, things we disagree on. One thing she is adamant about is a section near the beginning. My birth.

"I got to hold you first," she says. "When you were born. Just for a minute, I held you before the doctors took you. You got it backwards. I held you before they took you."

I don't get why this feels so important to her.

"I met you first," she continues. "Before all the other stuff with the doctors and their opinions and diagnoses, I held you in my arms. I met you."

IN HER BOOK *THE UNDYING,* ANNE BOYER WRITES ABOUT medical naming as a process that "takes information from our bodies and rearranges what came from inside of us into a system imposed from far away."

It is a way we are made unknowable to ourselves. In imposing a language foreign to us, we become foreign.

"A diagnosed person," Boyer says, "is liberated from what she once thought of as herself."

She, no longer the expert of her body. The experts, now the ones who name, the ones with the treatment, the ones ordained to fix.

Naming, of course, does not always lead to healing. Sometimes the naming is the sickness. Sometimes the named are forced to atone, fragment, bow to a societal conviction so insidious, so entrenched it feels like truth: that there is a limited way a body is allowed to be in the world.

But this is not truth. Or it doesn't have to be.

My biggest problem growing up was never the missing ear. It was the fixing.

THERE IS A PHOTO OF ME AS A CHILD, PROBABLY THREE years old. I'm standing on a patch of grass in a red T-shirt and red pants with a white racing stripe down the side. I'm wearing a red jacket and my arms are over my head in a V formation. The body of the jacket makes a cape behind me. My hair is short and disheveled. My lips, a pronounced lopsided triangle. A smile so big my eyes squint.

I used to hate this photo, this child's brazen unruliness. Now, I learn from her.

This girl is not afraid to make herself big. She doesn't worry about her crooked face, her missing ear. She is not burdened by the surgeries that will soon start, by the story that will be carved into her.

She is the closest I've come to knowing myself.

I have felt her sometimes glimmering inside me. Lowering my Halloween princess mask to take full breaths of fresh air, running away from my pediatrician pushing his fingers in fresh incisions, catching the jeweled calculator thrown at me. Saying no to Uncle Louie's plan for more surgeries—twice. Dancing with my eyes closed, feet stomping, hands flailing. Shoving my almost-boyfriend off me. Holding my belly through my miscarriage. Tenderly cleaning the ear.

She was there. She is here. Fighting for me.

She knows what I forgot when a surgeon pressed a plastic ear into me to show me his vision. She knows there is no such thing as an unfinished person.

This girl clad in red, arms to the sky as though to take flight, this is the girl I'm supposed to be. The one I am returning to.

Acknowledgments

I AM FOREVER GRATEFUL TO MANY FOLKS IN MY LIFE FOR the support in writing this book. It truly does take a village.

To my amazing agent, Sara Harowitz. Thank you for believing in this book, for your wisdom, your encouragement. For your hard work and dedication in getting this book into the right hands. I feel incredibly lucky to have a rock star like you in my corner.

Thank you to my team at Simon & Schuster. It has been thrilling to work with you! A big thank-you to Adrienne Kerr for your enthusiasm and passion for this book. For your kindness, compassion, and patience in guiding me through the process of publishing my first book. From the moment I met you, I knew the book was going to be in good hands.

Thank you to Michel Vrana for the gorgeous cover design and to Mary Beth Constant for your thoughtful and thorough copyedits.

To Amanda Lewis for making this book better with your insightful edits and generous feedback. I never knew an editing process could be so inspiring and collaborative. Thank you.

A big thank-you to friends and mentors who graciously supported and encouraged this project at its various stages: Mary Allen, Aeman Ansari, Katie Bannon, Dean Cooke, Tom Fulner, BK Lauren, Tanis McDonald, Margaret McPherson, Helen Porter, Ronan Sadler, Jacob Scheier, Sarah Sheard, and Sharon Singer. To Dafna Izenberg, who kept me standing with her continuous support of me in this project when I felt like falling to my knees. To Lise Sorokopud, who read the very first draft of this book many moons ago. Even though it was a complete mess, you told me you loved it. Thank you for being the first champion for this book. Thank you to the Holy Oak Café (RIP) and the beautiful friends who met me there week after week to write, drink coffee, and not talk much.

Thank you to Mr. Ed Walker, teacher extraordinaire. You were the first to encourage me to write my story. This book started on the lined pages you handed to me in the fifth grade. May you rest in peace.

Small sections of this book appeared in the *Humber Literary Review*, *The Malahat Review*, and Biblioasis's *Best Canadian Essays 2024*. Thank you to editors Eufemia Fantetti, Iain Higgins, and Marcello Di Cintio, respectively.

Thank you to the Iowa Summer Writing Festival. My seven summers in Iowa were some of my favorite summers ever. Thank you to Sage Hill and to all the wonderful friends I made there. Thank you to Artscape Gibraltar Point, the Humber School for Writers, the University of Toronto Summer Writing School, Takt Berlin Residency, and Residencia Corazón.

Thank you to the Toronto Arts Council and the Ontario Arts Council (and to Wolsak & Wynn and Second Story Press for their recommendations). I am truly grateful for your support.

Thank you to Sarah Manguso for inspiring the title with a line from your stunning book, *The Guardians*: "It must be very beautiful to be finished."

To Lucy Grealy: I found a home in your words. Thank you.

A big thank-you to my family for all the support throughout the years.

To my husband, James Braun, thank you for your unwavering

support and encouragement in getting to the finish line with this book. Thank you for the cuddles and pep talks. I appreciate you more than words can express.

To my mother, Susan Gies, thank you for holding my hand through the hard times, for talking through painful memories with me to help me piece together the fuzzy parts. For your wisdom and insight. For sleeping beside me every night in the hospital. To my father, Ken Gies, thank you for your continuous encouragement and faith in me, for the optimism and passion you bring to everything you do. I couldn't have asked for better parents; I'm incredibly grateful for you both.

To my friends on Ward 7C: We made our own little community under the sharp hospital fluorescents, and although our friendships were brief, they will be a part of me forever. Thank you. I think of you often.

Works Consulted

Boyer, Anne. *The Undying*. Farrar, Straus and Giroux, 2019.

Cole-Adams, Kate. *Anesthesia: The Gift of Oblivion and the Mystery of Consciousness*. Counterpoint, 2017.

Covino, Deborah Caslav. *Amending the Abject Body*. SUNY Press, 2004.

Dickinson, Emily. "The Saddest Noise, the Sweetest Noise." In *The Complete Poems of Emily Dickinson*, edited by Thomas H. Johnson. Back Bay Books, 1976.

Field, Thalia. *Point and Line*. Penguin Books Canada, 2000.

Flaum Hall, Michelle, and Scott E. Hall. *Managing the Psychological Impact of Medical Trauma: A Guide for Mental Health and Health Care Professionals*. Springer Publishing Company, 2017.

Flaum Hall, Michelle, and Scott E. Hall. "When Treatment Becomes Trauma: Defining, Preventing, and Transforming Medical Trauma." *VISTAS Online*, 2013. https://www.counseling.org/docs/default-source/vistas/when-treatment-becomes-trauma-defining-preventing-.pdf.

Frank, Arthur. "Emily's Scars: Surgical Shapings, Technoluxe, and Bioethics." *Hastings Center Report* 34, no. 2 (2004): 18–29.

Gilman, Sander L. *Making the Body Beautiful: A Cultural History of Aesthetic Surgery*. Princeton University Press, 2000.

Grealy, Lucy. *Autobiography of a Face*. Houghton Mifflin Company, 1994.

Grealy, Lucy. Interview by Charlie Rose, *Charlie Rose*, November 16, 1994. https://charlierose.com/videos/3763.

Hawthorne, Nathaniel. "The Birthmark." In *Hawthorne's Short Stories*, edited by Newton Arvin, 177–193. Vintage Books, 2011.

Works Consulted

King, Stephen. *Carrie*. Doubleday, 1974.

Lorde, Audre. *The Cancer Journals: Special Edition*. Aunt Lute Books, 1997.

NDiaye, Marie. *Self-Portrait in Green*. Translated by Jordan Stump. Two Lines Press, 2014.

Scarry, Elaine. *The Body in Pain: The Making and Unmaking of the World*. Oxford University Press, 1985.

Talley, Heather Laine. *Saving Face: Disfigurement and the Politics of Appearance*. New York University Press, 2014.

Williams, Terry Tempest. *When Women Were Birds*. Sarah Crichton Books, 2012.

Zwartjes, Arianne. *Detailing Trauma: A Poetic Anatomy*. University of Iowa Press, 2012.

KATE GIES is a writer and educator living in Toronto. She teaches creative nonfiction and expressive arts at George Brown College. Her fiction, nonfiction, and poetry have been published in *The Malahat Review*, *Humber Literary Review*, *Hobart*, *Minola Review*, and *The Conium Review*. She was a finalist for the CBC Nonfiction Prize, and her essay "Foreign Bodies" (excerpted from *It Must Be Beautiful to Be Finished*) was included in the *Best Canadian Essays* anthology.